DESIGN OF 3D INTEGRATED CIRCUITS AND SYSTEMS

Devices, Circuits, and Systems

Series Editor
Krzysztof Iniewski
CMOS Emerging Technologies Research Inc.,
Vancouver, British Columbia, Canada

FORTHCOMING TITLES:

Power Management Integrated Circuits and Technologies
Mona M. Hella and Patrick Mercier

Reconfigurable Logic: Architecture, Tools, and Applications
Pierre-Emmanuel Gaillardon

Radio Frequency Integrated Circuit Design
Sebastian Magierowski

Semiconductor Radiation Detection Materials and Their Applications
Salah Awadalla

Soft Errors: From Particles to Circuits
Jean-Luc Autran and Daniela Munteanu

VLSI: Circuits for Emerging Applications
Tomasz Wojcicki and Krzysztof Iniewski

Wireless Transceiver Circuits: System Perspectives and Design Aspects
Woogeun Rhee and Krzysztof Iniewski

DESIGN OF 3D INTEGRATED CIRCUITS AND SYSTEMS

EDITED BY
Rohit Sharma
Indian Institute of Technology Ropar
Punjab, India

Krzysztof Iniewski MANAGING EDITOR
CMOS Emerging Technologies Research Inc.
Vancouver, British Columbia, Canada

FOREWORD BY Sung Kyu Lim

CRC Press
Taylor & Francis Group
Boca Raton London New York

CRC Press is an imprint of the
Taylor & Francis Group, an **informa** business

CRC Press
Taylor & Francis Group
6000 Broken Sound Parkway NW, Suite 300
Boca Raton, FL 33487-2742

First issued in paperback 2020

© 2015 by Taylor & Francis Group, LLC
CRC Press is an imprint of Taylor & Francis Group, an Informa business

No claim to original U.S. Government works

ISBN-13: 978-1-4665-8940-7 (hbk)
ISBN-13: 978-0-367-65592-1 (pbk)

Library of Congress Cataloging-in-Publication Data

Design of 3D integrated circuits and systems / editor, Rohit Sharma.
    pages cm. -- (Devices, circuits, and systems)
    Includes bibliographical references and index.
    ISBN 978-1-4665-8940-7
    1. Three-dimensional integrated circuits. I. Sharma, Rohit, editor.

TK7874.893.D48 2014
621.3815--dc23
                                                                2014021624

Visit the Taylor & Francis Web site at
http://www.taylorandfrancis.com

and the CRC Press Web site at
http://www.crcpress.com

Contents

Preface

Traditional scaling of devices and circuits in line with Moore's law cannot be achieved by using the planar design approach alone. Due to increased challenges in maintaining technology advancements, alternative methods to achieve enhanced system-level performance have been envisioned. 3D integration of integrated circuits and systems has become imperative for the future growth of the semiconductor industry. The topic of 3D integration requires greater understanding of several interconnected circuits and systems stacked over each other. While this approach of vertical growth profoundly increases the system functionality, it also exponentially increases the design complexity. Newer processes are required for vertical interconnection of these stacked subsystems. Along with design of 3D integrated circuits (ICs) and systems, recent advances in interposer technology have resulted in the use of through-silicon vias (TSVs) with an aim to reduce power consumption and increase speed and functionality. The increased functionality and complexity in 3D ICs and systems necessitate a coherent design approach that takes into account distinct yet interrelated design and fabrication challenges. These include the use of interposer technology and the role of TSVs that require new processes. This also means that traditional materials should be replaced with novel materials to meet the higher speed and lower power budget requirements. Further, we need to address the issue of thermal reliability that has been initially underestimated. Traditional cooling techniques need to be revisited in view of the vertical integration. All in all, there is a greater need for synergy between these design aspects.

This present text on design of 3D ICs and systems is an attempt to focus on some of the aspects of 3D integration, including 3D circuit and system design, new processes and simulation techniques, alternative communication schemes for 3D circuits and systems, application of novel materials for 3D systems, and understanding the thermal challenges to restrict power dissipation and improve performance of 3D systems. The book aims to provide a detailed overview of these areas and explain their impact on the efficiency of 3D systems. Our objective has been to present a comprehensive source of literature of major design and fabrication issues related to 3D integration. We feel that this book will be a useful resource for researchers, industry designers, and graduate students initiated into this area. This text is composed of nine contributed chapters by experts from the industry as well as academia.

Chapter 1 provides an overview of the various processes involved in 3D chip stacking technology, including the formation of TSV and micro-bumps, wafer thinning, backside processing, bonding, and underfill encapsulation. In addition to TSV, 3D integration using various bonding technologies is also presented in this chapter. It also illustrates different 3D integration approaches, such as die-to-die, die-to-wafer, and wafer-to-wafer, and describes the best die-to-wafer technology for high yield and maximum manufacturing throughput. Some experimental results of chip-level 3D integration schemes tested at IBM are also presented.

Chapter 2 reports advanced complementary metal oxide semiconductor (CMOS) integration for 3D ICs with some case studies. It presents alternative wafer-level three-dimensional (3D) technology platforms with emphasis on dielectric adhesive bonding using benzocyclobute. Digital applications such as microprocessors and application-specific integrated circuits (ASICs) are discussed, with emphasis on performance prediction of static random access memory (SRAM) stacks for high-density memory. Analog/mixed-signal applications are also presented, with emphasis on wireless transceivers for software radios. Unique system architectures enabled by wafer-level 3D integration are reported, with emphasis on both (1) point-of-load (PoL) DC-DC converters for power delivery with wide control bandwidth and (2) a novel system architecture using optical clocking for synchronous logic without latency. Finally, technology and system drivers that need to be addressed before wafer-level 3D technology platforms are implemented for high-volume manufacturing are projected.

With continuous scaling, on-die integration of many components, including graphics processing units (GPUs) with central processing units (CPUs), has emerged. It has brought many challenges that arise from integrating disparate devices/architectures, starting from overall system architecture, software tools, programming and memory models, interconnect design, power and performance, transistor requirements, and process-related constraints. Chapter 3 provides insight into the implementation, benefits and problems, current solutions, and future challenges of systems having CPUs and GPUs on the same chip.

Chapter 4 is focused on thermal challenges of 3D ICs. Computer-aided design (CAD) tools and flows for modeling the dynamic interaction between electrical and thermal phenomenons, i.e., electrothermal simulation, are essential for successfully designing 3D ICs. This chapter describes an approach for system-level electrothermal simulation of 3D ICs, which allows us to study the trade-offs between various design choices for 3D integration early in the design phase. A case study comparing 2D and 3D implementations of a quad-core chip multiprocessor is presented in this chapter.

In Chapter 5, latest improvements in three major fields of thermal management for multiprocessor systems-on-chip (MPSoCs) have been described. Mathematical models related to the heat transfer model of the MPSoC, with special emphasis to the way the system cools down and the heat propagates inside the MPSoC, are presented. A liquid cooling model of a 3D-MPSoC is also provided. Detailed thermal profile estimation of the MPSoC structure is presented to achieve a temperature estimation by using thermal sensors placed in specific locations on the MPSoC. Also, this chapter describes a distributed thermal management policy suitable for liquid cooling technologies. The policy manages MPSoC working frequencies and micro-cooling systems to reach their goals in the most effective possible way and consuming the lowest possible amount of resources.

Chapter 6 presents a detailed overview of the various technologies available for interlayer communication in 3D networks-on-chip (NoCs). These include conventional copper-based TSVs, optical and nanophotonics, and wireless communication using inductive coupling. While metal interconnects are by far the most common choice, emerging interconnect technologies like nanophotonics, wireless, optical, and carbon-based interconnects are extensively researched as other alternatives. A qualitative comparison between some of the above communication technologies highlighting their pros and cons is presented in this chapter.

The concept of inductive coupling proposed above is further elaborated by authors in Chapter 7. ThruChip Interface (TCI) is a low-cost wireless version of TSVs. TCI wirelessly communicates over 3D stacked chips through inductive coupling between on-chip coils. TCI can provide competitive communication performance to TSV, even though it utilizes a wireless channel. Fusion combination between the inductive coupling channel and legacy wireline circuit techniques can further enhance communication performance. This chapter discusses some practical applications, such as NAND Flash memory stacking, and emerging applications, such as a noncontact memory card and a permanent memory system using TCI.

Chapter 8 presents the fabrication and modeling approach for Cu- and carbon nanotube (CNT)-based through-silicon vias (TSVs). In that, the fabrication and modeling of Cu-based TSVs along with their limitations are discussed. Based on the geometry and physical configurations, equivalent electrical models for Cu- and SWCNT bundle-based TSVs are presented. Cu-based TSVs suffer from performance limitations, including increased resistivity due to grain boundary scattering, surface scattering, and the presence of a highly diffusive barrier layer. These limitations can be removed using CNTs that result in effective via designing as well as offer higher mechanical and thermal stability, higher conductivity, and larger current-carrying capability.

Power dissipation during testing has emerged as a new threat to the quality and costs of large-scale integrated (LSI) testing, especially for low-power circuits. To tackle this serious problem of excessive test power, effective and efficient low-power testing must be conducted through sophisticated test power analysis and an optimal combination of test power reduction techniques. Chapter 9 provides the readers basic information about LSI testing and familiarizes them with state-of-the-art low-power testing solutions for 2D and 3D circuits and systems.

Overall, we have made an attempt to include various design challenges that are encountered in 3D integration in this book. We have also tried to include equitable contributions from academia and industry. We hope that our readers will find these topics interesting and helpful in understanding this fascinating area. We look forward to hearing from you.

Rohit Sharma
Krzysztof Iniewski

Foreword

Lately we are seeing an increasing number of articles on the end of Moore's law. As of spring 2014, the finFET technologies from Intel, TSMC, and Samsung were promising a successful (and profitable) landing onto the 16/14 nm generation with mainstream products scheduled in fall 2014. Meanwhile, there is a growing concern on whether any geometry smaller than 16/14 nm such as 10, 7, and 5 nm will continue to be reliable and profitable.

Many in the semiconductor industry have believed and demonstrated that 3D ICs using through-silicon-via (TSV) is the leading contender to continue Moore's law and promise a steady increase in the number of IC devices for various applications in mobile and cloud computing for consumer, biomedical, and environmental electronics. This is backed by several product announcements using 3D IC technologies from major vendors including Samsung, Nvidia, Xillinx, Qualcomm, etc. TSMC and GlobalFoundries have announced their readiness of low-cost 3D IC manufacturing. Lastly, major EDA vendors have begun to offer much-needed CAD tools for the designers. It is now very hard to deny that the era of 3D IC and TSV is here.

As an author of a book on the same topic, I am thoroughly enthusiastic about *Design of 3D Integrated Circuits and Systems* edited by my dear colleagues Profs. Rohit Sharma and Krzysztof Iniewski. This book offers a wide spectrum of burning issues and their solutions in modeling and design of TSV-based 3D ICs written by the world's leading experts. The book starts with two chapters on up-to-date coverage of manufacturing technologies for 3D ICs by IBM and RPI, two major players in this area. The issues in manufacturing always have a profound impact on every 3D IC designer including me. In addition to CPU+memory stacking, the integration of GPU and CPU has emerged as another killer application for 3D IC. The authors from Rambus and Intel provide an industry perspective on this topic, which I believe will inspire stimulating debates among architects, circuit designers, and technology developers.

Thermal issues continue to threaten 3D IC reliability, and the two chapters on this topic written by the two leading groups at NCSU and EPFL are highly relevant, informative, and innovative. The book then moves on to discuss the die-to-die communication using inductive coupling, an alternative 3D IC technology that does not utilize TSVs. This technology, studied by several groups in Asia including IIT Ropar, SNU, Kobe University, and Keio University, overcomes the electro-mechanical reliability problems associated with TSVs while offering ultrahigh speed and bandwidth inter-die communication with low noise. Next, readers will find the comparison between TSVs made with copper versus carbon nano-tubes conducted by researchers from IIT Roorkee forward-looking and intriguing. The book concludes with a chapter on 3D IC testing, which is another burning issue that hinders commercial application of 3D ICs today, written by co-authors from Mentor, Kyushu and Tsinghua University.

As a researcher in 3D IC modeling and design, I find this book timely and invaluable. It offers a balanced view between academia and industry, which adds an additional value to the readers. The 3D IC R&D community truly needs more books like this to keep it engaged, informed, focused, and vibrant. It is thus my great pleasure to endorse this book.

Sung Kyu Lim
Atlanta, Georgia, USA

About the Editors

Rohit Sharma is on the faculty of electrical engineering at the Indian Institute of Technology Ropar since July 2012, prior to which worked for one year as a post-doctoral researcher in the Design Automation Lab at Seoul National University. During that time, he was involved in development of novel communication schemes for 3D Network-on-Chips (NoCs). In 2011, he moved to the Interconnect Focus Center (IFC) at Georgia Institute of Technology, where he worked for one and a half years as a post-doctoral researcher and was involved in design of ultra-low loss air-clad interconnects for chip-chip applications.

Dr. Sharma's research interests include design of high-speed chip-chip and 3D interconnects, and communication schemes for multi-core architecture. He authored *Compact Models and Measurement Techniques for High-Speed Interconnects* (Springer, 2012) and is the author or co-author of more than 50 journal publications and conference proceedings, one book chapter, two patents/copyrights, and several invited talks.

Dr. Sharma is a recipient of the Brain Korea Research Fellowship (2010), the Indo-US Research Fellowship (2011), and Best Paper Awards in ASQED 2010 and GIT 2011 conferences. He has worked as a referee for several journals including *IEEE TEMC, IET MAP, Micro and Nano Letters, Journal of Supercomputing,* and *PIER*, and has served as a committee member/session co-chair for multiple international conferences. He is a member of the IEEE and ACM.

Krzysztof (Kris) Iniewski is managing R&D at Redlen Technologies Inc., a start-up company in Vancouver, Canada. Redlen's revolutionary production process for advanced semiconductor materials enables a new generation of more accurate, all-digital, radiation-based imaging solutions. He is also a President of CMOS Emerging Technologies (www.cmoset.com), an organization of high-tech events covering Communications, Microsystems, Optoelectronics, and Sensors.

In his career, Dr. Iniewski has held numerous faculty and management positions at University of Toronto, University of Alberta, Simon Fraser University, and PMC-Sierra Inc. He has published over 100 research papers in international journals and conferences and has written and edited several books for Wiley, IEEE Press, CRC Press, McGraw-Hill, Artech House, and Springer. He is also a frequent invited speaker and has consulted for multiple organizations internationally. Dr. Iniewski holds 18 international patents granted in USA, Canada, France, Germany, and Japan.

His personal goal is to contribute to healthy living and sustainability through innovative engineering solutions. In his leisure time Kris can be found hiking, sailing, skiing or biking in beautiful British Columbia.

He can be reached at kris.iniewski@gmail.com.

Contributors

Chinnakrishnan Ballapuram
Intel, Inc.
Folsom, California

Kiyoung Choi
Department of Electrical and Computer
    Engineering
Seoul National University
Seoul, Korea

W. Rhett Davis
North Carolina State University
Raleigh, North Carolina

Paul D. Franzon
North Carolina State University
Raleigh, North Carolina

Ronald J. Gutmann
Rensselaer Polytechnic Institute
Troy, New York

Jianchen Hu
North Carolina State University
Raleigh, North Carolina

Brajesh Kumar Kaushik
IIT Roorkee
Roorkee, Uttarkhand, India

Manoj Kumar Majumder
IIT Roorkee
Roorkee, Uttarkhand, India

Archana Kumari
IIT Roorkee
Roorkee, Uttarkhand, India

Tadahiro Kuroda
Keio University
Yokohama, Japan

Xijiang Lin
Mentor Graphics Corp.
Wilsonville, Oregon

Jian-Qiang Lu
Rensselaer Polytechnic Institute
Troy, New York

Noriyuki Miura
Kobe University
Kobe, Japan

Shivam Priyadarshi
North Carolina State University
Raleigh, North Carolina

Katsuyuki Sakuma
IBM T.J. Watson Research Center
Yorktown Heights, New York

Deepak C. Sekar
Rambus, Inc.
Sunnyvale, California

Rohit Sharma
Department of Electrical Engineering
Indian Institute of Technology Ropar
Rupnagar, India

Michael B. Steer
North Carolina State University
Raleigh, North Carolina

Xiaoqing Wen
Kyushu Institute of Technology
Kitakyushu, Japan

Dong Xiang
Tsinghua University
Beijing, China

Francesco Zanini
École Polytechnique Fédérale de
    Lausanne
Lausanne, Switzerland

1 3D Integration Technology with TSV and IMC Bonding

Katsuyuki Sakuma
IBM T.J. Watson Research Center

CONTENTS

1.1 INTRODUCTION

In accord with Moore's law, the refinement of integrated circuit technology has been doubling the number of devices in a given chip area every 2 years. However, conventional device scaling is approaching its physical limits, and physics-based constraints will force many changes in materials, processes, and device structures as the industry moves down to 22 nm and smaller designs. There are problems such as increasing capital investments (such as for better lithography tools), while the technical problems such as increasing interconnect wire delays and gate leakage currents in the integrated circuits are becoming insurmountable barriers to sustaining past rates of performance growth.

The most important problem is that the average interconnect's length now has a significant impact on system performance. Gate delays used to be the major concern in microprocessors. In addition, the signal propagation delays in the very-large-scale integrated (VLSI) circuit's interconnections are becoming a more serious technical problem than ever before. Figure 1.1(a) shows the interconnect delay problem [2]. The interconnect delays caused by parasitic resistance and capacitance have become the predominant contributors to total delay time. The interconnect delays caused approximately 75% of the overall delay in the 90 nm generation. Replacing aluminum and SiO_2 with copper and low-k interlevel dielectrics (ILD) for multilevel metallization has helped reduce this effect. However, technological solutions based on copper and low-k dielectrics have only slowed the increase of the delay times and are not fundamental solutions, since there is no alternative interconnect metal to replace copper. With further miniaturization, interconnect delays will become increasingly dominant parts of the total delays.

Second, power dissipation in current VLSI is also becoming a large technical problem. Power dissipation is becoming another barrier to sustaining the rate of performance improvement in the future. As the channels are becoming shorter, the gate leakage current, which is controlled by the physical thickness of the oxide, is increasing with the device scaling. Figure 1.1(b) shows the relationship between technology node and normalized power [4]. Historically, dynamic power was dominant for total power consumption. At smaller feature size, gate leakage current is contributing more significantly to the passive power. Passive power levels are

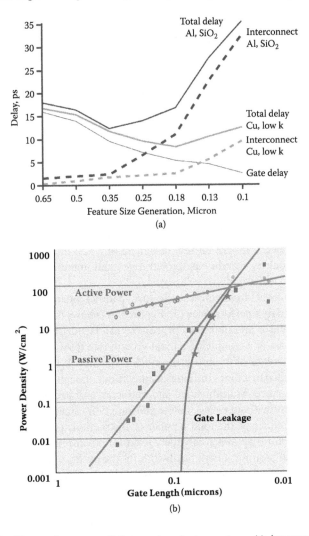

FIGURE 1.1 The performance of the semiconductor system: (a) interconnection delay problem [2] and (b) relationships between technology node and normalized power [4, 108]. (Courtesy of ITRS.)

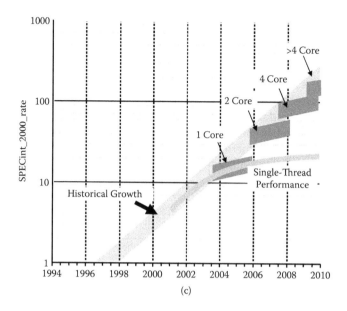

FIGURE 1.1 (Continued)   The performance of the semiconductor system: (c) performance growth vs. multicore processors [4].

approaching the active power and are beginning to dominate the power consumption and wasting much of the power budget in high-performance microprocessors. Thus, optimizing not only the active power but also the static power consumption is a major concern for the current designers of VLSI chips. Consequently, it is increasingly difficult to sustain past growth rates and manufacture each successive generation of chip technology.

Third, system-level architectures now use parallel operations to try to meet requirements for higher performance [3]. Figure 1.1(c) shows the relationship between performance increases and multicore processors [4]. Historically, single-thread performance has grown at a compound annual growth rate (CAGR) of about 45%, consistent with the increase of the processor clock frequency. It has become difficult to make single-threaded machines run significantly faster, though there are still demands for higher performance. Multithreaded and multicore architectures can improve the processing capacity without excess power consumption. Since each die runs in parallel at reduced frequency, the total number of instructions per second continues to rise. This architectural approach is widely used for advanced servers [5]. However, a multicore system requires a high bandwidth for the cache, and there are geometric constraints on the parallelism in two dimensions [6, 7]. This is because the lengths of the wires between the cores and the caches increase as more cores are added in 2D, and therefore the wiring delays are become problematic in the complete systems.

One of the potential solutions for these problems is 3D integration. 3D integration technology is expected to reduce interconnect delays and dramatically increase chip performance and package density, since it can address the serious interconnection problems while offering integrated functions for higher performance. For maximum use of the potential of multicore architectures, 3D integration of memory and processors is a promising solution. A 3D integration technology can provide a large number of signal paths between the core and the cache. Such 3D integration technology has been receiving increasing amounts of attention, not only from VLSI researchers and industry experts, but also from packaging researchers, because such technology is also useful for new packaging applications. In addition to processors and caches, memories, radio frequency (RF) devices, sensors, microelectromechanical systems (MEMS), and bio-microsystems can be integrated as packaged applications [8–10]. 3D integration offers many benefits for future VLSI chips and packaging.

In a widely used approach, the thinned chips are stacked directly onto other chips and connected electrically by using a 3D chip integration technology. Key technologies for 3D chip integration include forming high-aspect-ratio via in silicon, filling the TSV with conducting material, forming micro-bumps, wafer thinning, precise chip alignment, and bonding. The advantages of 3D integration technology are discussed in more detail in the next section. Approaches to 3D integration for high-performance VLSI are also reviewed. This chapter describes the 3D chip integration process development and the results of characterization of die-to-wafer 3D integration.

1.2 3D INTEGRATION TECHNOLOGY

3D integration spans from the transistor buildup, silicon-on-insulator (SOI) device-stacking level to the bulk thin-silicon die-stacking level, and the silicon packaging level. Different levels of 3D integration were previously studied and reported on by companies, consortia, and universities. This section reviews the various kinds of 3D integration technology, including their advantages and limitations.

1.2.1 Advantages of 3D Chip Integration

3D integration with TSV and micro-bumps is an attractive technology to meet the future performance requirements of integrated circuits. Compared to conventional 2D, the total footprint can be reduced when the functional device components are vertically stacked on top of each other by using 3D integration technology. For example, in layout designs of 2D and 3D inverters with fan-ins equal to 1, large areal gain such as 30% can be achieved with 3D, as shown by Ieong [11] in 2003. By stacking chips with 22 nm features rather than shrinking the device dimensions, there are advantages in power consumption, bandwidth, delay, noise, and packaging density.

3D integration with vertical integration technology can shorten the interconnections between functional blocks. The shorter interconnect wires will decrease both the average parasitic load capacitance and the resistance. Therefore, 3D integration will improve the wire efficiency and significantly reduce noise and the total active power [12]. High-density vertical interconnections between stacked layers can also be achieved by 3D integration with vertical interconnections. This provides extremely high bandwidth values and dramatically decreases the interconnection delays. Signal propagation delays and interconnection-associated parasitic capacitance and inductance are reduced by the shorter reduction of interconnection length using vertical interconnections [13–15].

1.2.2   LIMITATIONS OF 3D CHIP INTEGRATION

The performance and packaging density of electronic systems can be improved with 3D chip integration technology. However, there are several limitations, such as thermal management and design complexity, that must be considered when using a 3D integration technology. These problems are briefly outlined in the following subsections.

1.2.2.1   Thermal Management

One of the main challenges facing 3D integration is thermal management to satisfy the required performance and reliability targets [16]. In changing from 2D to 3D, the chip footprint decreases and the heat generation per unit of surface area increases. From a packaging perspective, a small footprint is better for smaller devices and products. However, the contact area in 3D stacked ICs is limited compared to traditional 2D circuits, making the heat dissipation more challenging. This thermal problem can be reduced when chip stacking uses lower-power circuits such as dynamic random access memories (DRAMs). Typical DRAM chips have power densities of only about 0.01 W/mm^2, which is much smaller than the 2 W/mm^2 power densities of the hotspots of some microprocessors [3]. Since the DRAMs need little power, stacking DRAMs directly on the processors can be a very attractive 3D application.

Responding to the demands for higher-performance systems, circuit density in 3D integration will increase and these higher-density circuits will increase the power density (W/cm^2) of the systems. The upper layers may prevent the cooling of the lower layers that rely on heat sinks, since the heat sinks are traditionally attached to the top surfaces of the chip packages. Therefore, thermal management is increasingly important in the design of high-performance stacked devices.

Other problems are the thermal stress and warpage encountered between different structures and surface boundaries, such as silicon or organic-based electronic modules made of different materials [17]. Thermal and mechanical analyses are needed to find the best electrical performance with high product reliability when using 3D integration.

1.2.2.2   Design Complexity

The new design paradigms from conventional 2D to 3D designs require changes in design methodologies [18]. Since conventional design tools are based on the algorithms for 2D circuits and direct extensions of 2D approaches, they are unable to solve the 3D design problems [19]. The problems of 3D design are related to topological arrangements of blocks, and therefore the physical design tools for 3D address unsolved problems related to global routing, standard cell placement, and floor planning [20]. Hotspot analysis engines are also needed as design tools. Designers need to consider and evaluate the thermal impacts on the circuits to control the hotspots. Both vertical and lateral heat transfer paths must be taken into account. As 3D integration becomes more sophisticated with such features as heterogeneous layers, the physical design problems have to be addressed and solved.

1.2.3 VARIOUS KINDS OF 3D TECHNOLOGY

Figure 1.2(a) illustrates chip integration approaches with increasing levels of integration. The physical parameters of the vertical interconnect, including the via sizes, pitches, and heights, are different for each integration level. The 3D stacks that are now emerging involve package-on-package (PoP) [21] and system-in-package (SiP), including wire bonded chip stacks at the silicon packaging level.

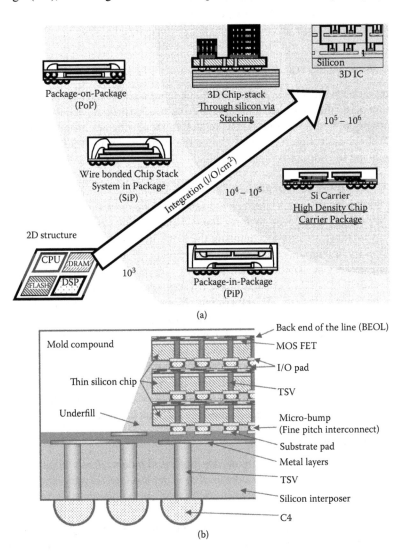

FIGURE 1.2 3D chip stack: (a) Emerging 3D silicon integration. Relative comparisons of I/O densities for 2D and 3D, including 3D silicon packaging, 3D chip stacks, and 3D ICs [59]. (b) Schematic of 3D chip stacking on a silicon interposer using the face-to-back approach [34]. (From Sakuma et al., in *Proceedings of the 57th Electronic Components and Technology Conference (ECTC)*, 2007, pp. 627–632. Copyright © 2007 IEEE.)

PoP-type integration packaging technology needs ball grid array packages for stacking each layer to save space. For SiP applications, multifunction component dies such as logic and memory can be stacked and connected by wire bonding interconnects in a single miniaturized package. While these approaches allow known-good device testing prior to stacking devices, disadvantages include long connection lengths and limited connections between chips. In order to overcome these wiring connectivity problems, 3D chip integration technology using TSV and micro-bumps is attractive because this offers a way to solve the interconnection problems while also offering integrated functions for higher performance [22–24].

1.2.4 Approaches for 3D Integration

3D integration with vertical interconnections is an attractive technology to satisfy the future performance needs of integrated circuits. Two primary schemes, top-down and bottom-up, and several variations related to these schemes have been proposed for 3D integration. Figure 1.3(a-b) shows the comparison between the top-down approach and the bottom-up approach.

1.2.4.1 Bottom-Up Approach

In the bottom-up approach, devices are fabricated on crystalline layers above the first layer of active devices to form the 3D integration structure. An additional layer can be fabricated on top by using the same methods and process sequences. Therefore, the interconnections and layering processes are performed sequentially. Different processing solutions such as beam recrystallization, selective epitaxial growth (SEG) [25–27], and solid phase crystallization [28–31] have been used in bottom-up approaches to 3D integration. Figure 1.3(c–d) shows schematics of multiple silicon layers using silicon epitaxial growth. The major problem is the high processing temperature (~1,000°C) for silicon epitaxial growth, and this significantly affects the lower layers of the devices, especially the layers with metallization. Lower temperature processing is needed for circuit integration. It is also difficult to control variations in grain sizes. Unless the grain variation can be controlled, building high-performance 3D integrated devices is difficult. This approach still needs improvements before it can be used for high-performance 3D integration.

1.2.4.2 Top-Down Approach

In the top-down approach, 2D circuits are fabricated independently in parallel, then aligned and bonded together at the die level or wafer level for 3D integration [22, 32]. The vertical interconnections between each layer can be fabricated before or after each layer is stacked. Depending on the target via size and pitch, either a bulk or an SOI-based complementary metal oxide semiconductor (CMOS) can be used for stacking, considering the various performance requirements. In contrast to the bottom-up approach, the thermal constraints when making the circuits are not a major concern. Each approach has advantages and limitations. However, we are focusing on the top-down approach because of thermal budget considerations and the process yield. Top-down 3D integration avoids some of the serious problems, such as variations in the thermal history affecting the device fabrication in each layer.

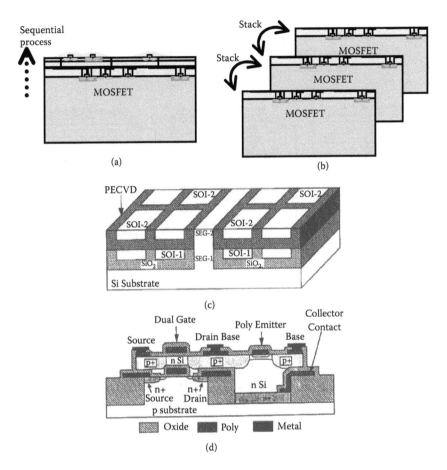

FIGURE 1.3 Different approaches for 3D integration: (a) bottom-up approach, (b) top-down approach. (c) multiple silicon layers using silicon epitaxial growth [26], (d) cross section of the 3D CMOS cell using selective epitaxial growth of silicon [109].

1.2.5 Key Enabling Technologies for 3D Chip Integration

A cross-sectional structure of face-to-back 3D chip stacking on a silicon interposer is shown in Figure 1.2(b). Each circuit layer with a different function and a silicon interposer is electrically connected through the vertical interconnections, including TSV and high-density low-volume solder interconnects. Several methods to achieve 3D integration using chip- or wafer-stacking technologies have been proposed [22, 33].

Figure 1.4 shows an example of the processing flow for 3D chip integration by using a die-to-wafer top-down approach. Some of the key technologies needed to enable 3D chip stacking include wafer thinning, the formation of TSV, the formation of micro-bumps, chip alignment and bonding, underfilling, and packaging. Each key process is described here. Table 1.1 lists the processes for 3D chip integration and their processing challenges.

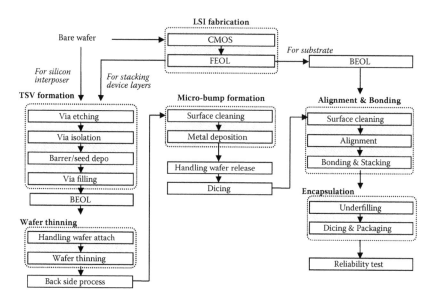

FIGURE 1.4   An example of process flow for a 3D chip integration [1].

TABLE 1.1
List of Processes for 3D Chip Integration and Processing Issues

Process	Challenges
Wafer thinning	• Surface quality
	• No substrate damage
	• Heat-resistant wafer support system
TSV formation	• High-quality via etching and filling processes for high throughput
	• Void-free via filling
	• Reliable conductive materials for reduced contamination, low CTE, and highly conductive materials
Micro-bump formation	• Processes for high density
	• Materials for high reliability, low resistance, and high thermal stability
Alignment and bonding	• High-precision alignment
	• High-reliability bonding
	• High throughput on stacking
	• High yields on stacking
Encapsulation	• Higher reliability underfill materials with a specified CTE
	• Void-free processes

Wafer thinning: After the front side processing is finished, the wafer is flipped over and thinned from the backside. During the thinning process, the devices and circuits on the front side must be protected from mechanical damage and chemical corrosion [35]. Reducing damage and residual stress in the polished surface is necessary to realize 3D integration for highly reliable devices. Minimizing silicon warpage and a smooth surface are required for further backside processing. A wafer support system is also required to handle the ultra-thin silicon substrates for the backside processing that forms the contact pads and micro-bumps [36]. After the backside processing, the handler is released, and the wafer surface is cleaned using plasma etching. Metal contamination from the backside surface caused by thinning process must be taken into account [110].

TSV formation: Through-silicon via (TSV), vertical interconnections in the silicon, allow for the shortest interconnections between the chips or wafers stacked for 3D integration [37]. The TSV should also assist in the heat dissipation. For the TSV formation, deep via holes are etched anisotropically into the silicon substrate, in most situations by using inductive coupled plasma reactive ion etching (ICP RIE) from either the front or back side. Next, the vias are isolated with dielectrics and filled with conducting material. The various options are discussed in Section 1.3. Research is still continuing into the most appropriate materials and processes for TSV.

Micro-bump formation: In addition to TSV, 3D integration requires bonding technology. Various kinds of bonding methods have been tested, such as oxide bonding, adhesive bonding, metal bonding, and combinations, and are discussed in Section 1.4. A bonding method can also be chosen based on the target interconnection pitch and tolerance of the fabrication process for 3D applications. For creating solder bumps, several bump-forming methods such as plating, evaporation [75], stencil printing, and controlled collapse chip connection/new process (C4NP) [76] have been proposed. Bonding at a low temperature is desirable to reduce stress and deformation during the bonding process.

Alignment and bonding: A high-precision chip alignment and bonding system is required for the fabrication of a variety of 3D integrated devices [38]. Alignment, positioning, and reflow solder or direct metal bonding are among the sequential processes for chip assembly. Suitable pressures and temperatures are used for solder bonding. The bonding alignment must be precisely controlled for accurate fine-pitch interconnections in 3D integration. 3D integration approaches based on die-to-die, die-to-wafer, and wafer-to-wafer are discussed in Section 1.5.

Encapsulation: After the chips are stacked, they can be underfilled and packaged. The micro-bumps need to be encapsulated to resist fatigue for improved reliability and performance [39]. By encapsulating the bumps in each stacked layer, the robustness of the 3D integrated structure is improved. This means that the underfill material selection has a large impact on reliability. Most of the suitable assembly underfill materials include filler content, use filler particles of a set size, and have a specified coefficient of linear thermal expansion (CTE). The proper material should be considered and selected for small gap underfilling for 3D integration. Different underfill approaches are discussed further in Section 1.6.

1.3 TSV

For 3D system integration, the thinned chips are stacked directly onto other chips, and they are connected electrically by using vertical interconnections. One of the key technologies for 3D chip fabrication is how the vertical interconnections are formed, which includes TSV and bonding technologies. Companies, consortia, and universities are developing different kinds of structures and processes for TSV and bonding methods [22, 40]. TSV processing includes via etching, insulating the sides of the vias, and via filling by using conductive material. Depending on the TSV processing time during the overall VLSI wafer processing, there are different options, such as via-first, via-middle, or via-last. These technologies have been demonstrated at IBM and elsewhere. There are different technical approaches for the manufacturing processes, materials, and physical structures for producing TSVs and micro-bumps [41, 42].

1.3.1 Processing Flow for TSV

TSV processes are classified into three groups, via-first, via-middle, and via-last. Via-first includes the TSV manufacturing methods that make the TSV before the transistor fabrication. Via-middle is making TSV after front-end-of-the-line (FEOL) processing but before back-end-of-the-line (BEOL) processing. Via-last involves making TSV after the BEOL processing. The phosphorus-doped polysilicon is used to make the via-first TSV to avoid metal contamination problems [22]. Via-first TSVs have comparatively small aspect ratios of around 3–10 and diameters of around 1–5 μm. In contrast, via-last TSVs can be classified more precisely depending on the processing in which TSV is made after the BEOL processing. Via-last TSVs have aspect ratios of around 3–20 and diameters of around 10–50 μm. The advantage of via-last is that the device manufacture is possible with foundry companies that do not have TSV processing facilities.

1.3.2 Via Etching

It is necessary to form the via with high aspect ratios for the TSV manufacturing. In a typical process, deep silicon anisotropic vias are formed with a dry etching process using an inductively coupled plasma (ICP) system that includes a 13.56 MHz RF power source to supply plasma and another 13.56 MHz generator to provide a platen supply. The wafer is cooled using helium back cooling to maintain a constant wafer temperature during processing. Sulfur hexaflouride (SF_6) and octafluorocyclobutane (C_4F_8) gases are used as plasma sources for etching and passivation, respectively [43]. A schematic representation of a passivation layer deposition model and an etching mechanism model is shown in Figure 1.5(a). An etching cycle using SF_6 and a polymer deposition cycle using C_4F_8 for passivation are alternated every 5–15 s to form anisotropic via. The SF_6 is primarily responsible for the silicon etching, and the etching action relies on chemical reactions involving the ions and decomposed radical species caused by electron impact dissociation:

$$SF_6 + e^- \rightarrow S_xF_y^+ + S_xF_y^* + F^* + e^- \tag{1.1}$$

$$Si + F^* \rightarrow SiF_x \tag{1.2}$$

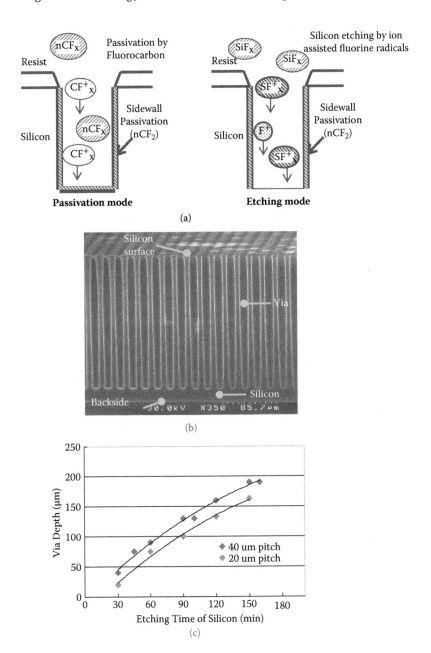

FIGURE 1.5 (a) Schematic representation of the passivating and etching mechanical model. (b) SEM cross-sectional view of the 20 μm pitch through-silicon vias before electroplating. (c) Correlation of via depth with etching time of silicon. (From Sakuma et al., *IEEE Trans. Electrical Electronic Eng.*, 4, 339–344 (2009). Copyright © 2009 IEEJ.)

In this environment, SiF_4 is a volatile gas. C_4F_8 is used in the passivation phase, and a polymer film is deposited on the sidewalls and the bottoms of the via by dissociation of the C_4F_8 [44]. Here are the reactions:

$$C_4F_8 + e^- \rightarrow CF_x^+ + CF_x^* + F^* + e^- \qquad (1.3)$$

$$CF_x^* \rightarrow nCF_2 \qquad (1.4)$$

$$nCF_2 + F^* \rightarrow CF_x^* \rightarrow CF_2 \qquad (1.5)$$

During the etching cycle, the passivation is preferentially removed from the bottom of the via by the ion bombardment, without etching into the sidewall regions.

By optimizing both the etching and passivation times in the Bosch processing [45], or the time-domain multiplexed (TDM) processing, in which the etching and passivation steps alternate, deep TSVs are created. Figure 1.5(b) shows a cross section scanning electron microscopy (SEM) of 20 μm pitch TSVs before the via filling process, and Figure 1.5(c) shows the relationship between the via depth and the etching time for silicon [46]. Photoresist was used as a mask layer during the silicon etching. The etching conditions had the coil power fixed at 600 W, with gas flow rates of 85 sccm for C_4F_8 and 130 sccm for SF_6. As shown in Figure 1.5(c), there are different etching rates for the dense and sparse regions of the patterns (between the 20 μm pitch and 40 μm pitch vias). The evidence indicates that etching species concentration at the bottom of the via decreases with increasing via depth and decreasing via pitch.

However, in the case of Bosch processing, corrugation of the sidewalls (scallops) occurs, and it becomes difficult to make the metal layer films uniform when the sidewalls are rough. Therefore, the method of forming the deep via in a non-Bosch process is being developed for manufacturing. In addition, problems based on micro-loading effect and notching effect occur depending on the target structure and processing conditions. The micro-loading effects mean that etching speed and depth are different, depending on the area and shape of the mask apertures. The notching effect occurs when there is an oxidation in the via bottoms due to accumulation of charge at the dielectric bottom layer, causing deflection of ions to the sidewalls at the via bottoms. The optimization of the etching condition is necessary so that such problems do not occur.

1.3.3 INSULATION LAYER

Vias are lined with a dielectric because TSVs filled with conductive metal need an insulation layer for electrical isolation from the surrounding bulk silicon. Thermal oxide processing could be used to grow SiO_2 in the range of 800–1,100°C for dielectric isolation, but such a high-temperature process is acceptable only for via-first processing. Amorphous SiO_2-based dielectric films can be deposited by chemical vapor deposition (CVD) in the range of 150–400°C, and this is suitable for via-middle and via-last methods, since temperatures below 400°C are safe for devices and metal wiring, and the CVD process does not require high temperatures. Another concern is the step coverage. For plasma-enhanced CVD (PECVD), the gas phase reactions are initiated by RF plasma instead of thermal energy. However, conventional PECVD has a low step coverage and isn't suited for high-aspect-ratio TSV structures. Subatmospheric CVD (SACVD) using tetraethylorthosilicate (TEOS) and O_3 can be used for the high-aspect-ratio vias to improve the step coverage of the SiO_2 film due to their high conformal deposition behavior. The step coverage for SACVD is better than for PECVD, and the sidewall roughness can be reduced and scallops can be reduced by using SACVD.

Low dielectric constants of the materials are required to lower capacitance of the wiring. For example, Interuniversity Microelectronics Centre (IMEC) reports that it is necessary to reduce the capacitance below 50 fF to achieve the same speed with 3D compared to 2D [47]. Small RC delays of the TSV are needed for improved device performance. Conventional SiO_2 is typically formed in the range of 100 nm–1 μm as the TSV dielectric, since the thicker ranges are difficult to obtain on the TSV sidewalls with good quality. Since thicker layers can be achieved by using polymers, the dielectric polymer insulation film is also suggested for electrical insulation that reduces the mechanical stress and lowers the TSV parasitic capacitance [48].

1.3.4 BARRIER AND ADHESION LAYER

Tungsten (W) and copper (Cu) are candidate conductive materials for via-middle and via-last TSVs. For Cu-filled TSV, barrier layer formation on TSV sidewall is needed to block the Cu diffusion. Ti, TiN, and Ta are widely used as the diffusion barrier materials in Al and Cu metallizations, and therefore they are deposited for TSV structures by CVD or physical vapor deposition (PVD) processes. Highly conformal barrier layers are needed for high-aspect-ratio vias. TiN is also used as an adhesion layer for tungsten CVD. In the case of high roughness of TSV scallops, it is challenging to deposit conformal oxide liner and barrier layer. If the coverage of the barrier layer is poor, Cu atoms (conductive material of TSV) may diffuse to silicon substrate during post-annealing process and cause severe degradation in device reliability [58, 114].

1.3.5 CONDUCTIVE MATERIALS FOR TSV

Different candidate metals such as Cu, tungsten, Ni, and doped polysilicon have been used as conductive materials in the vias. The mechanical characteristics of silicon and candidate TSV conductive materials are shown in Table 1.2.

Examples of TSV fabricated using different technologies are shown in Figure 1.6. The most common approach is to metalize the via with Cu, since Cu has low electrical resistance and is often used for wiring [49–51]. CVD or electroplating is used for the Cu via filling processing, and the selection depends on via size and conductive material. For Cu electroplating, seed layer deposition with physical vapor deposition (PVD) is needed to deposit a continuous Cu seed layer in high aspect ratio TSV structure. A high-quality seed layer is needed since insufficient seed layer coverage at the bottom of TSV causes bottom voids in the TSV. If a robust filling process is not used for the copper filling, voids may form within the via. In addition, the large

TABLE 1.2
Mechanical Characteristics of Silicon and Candidate TSV Conductive Materials

Material	Coefficient of Thermal Expansion @ 293K ($\times 10^{-6}$/K)	Melting Temperature (°C)	Resistivity @ 273K ($\mu\Omega$ cm)	Thermal Conductivity (W/m^{-1}·K^{-1})
Si	2.6	1,414		150
Cu	16.5	1,084	1.55	394
W	4.5	3,410	4.9	177
Ni	13.4	1,455	6.2	94

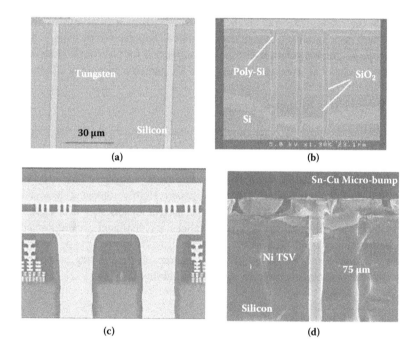

FIGURE 1.6 Examples of TSVs fabricated with different technologies: SEM images of (a) IBM's tungsten TSV [37], (b) Tohoku University's polysilicon TSV [53], (c) Cu TSV [111], (d) Ni TSV [46].

CTE disparity can affect the performance of the transistors and the Cu plug may move vertically, so-called Cu-pumping, (Figure 1.7) [58]. To avoid Cu-pumping effect, Cu must be annealed. Annealing could reduce the Cu extrusion, but induce several issues such as Cu voids formation and diffusion of Cu atoms. Because there is a significant mismatch in the CTE between Cu and bulk silicon, Cu TSV induces stress on its surroundings. In this case, a device prohibition area called keep-out zone (KOZ) is used, and must be included in the circuit layouts [52]. Also, the Cu

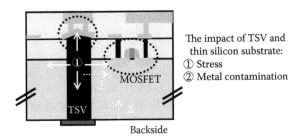

The impact of TSV and thin silicon substrate:
① Stress
② Metal contamination

Backside

FIGURE 1.7 The impact of TSV and thin silicon substrate. (From Sakuma, *J. Inst. Electrical Eng. Jpn.*, 131(1), 19–25 (2011) (in Japanese). Copyright © 2011 IEEJ.)

contamination problem needs to be handled, since Cu spreads to silicon at high temperatures and the contamination affects the characteristics of semiconductor devices (Figure 1.7) [58].

Doped polycrystalline silicon (poly-Si) via can be used instead of a Cu via, since poly-Si uses the same substance as the Si itself, and contamination of alien atoms into the silicon is avoided. However, the total processing time must be considered, depending on the via size and depth, and the resistance of such a via is extremely high compared to Cu or tungsten, even when the poly-Si is doped. In general, n^+ poly-Si is deposited by a low-pressure chemical vapor deposition (LPCVD) method using SiH_4 gas after the sidewall of the via is thermally oxidized. In order to decrease the resistance of the TSV interconnections, a phosphorus doping process is used [9, 53]. This is followed by an annealing process that diffuses the dopants into the entire vertical interconnection at 1,000°C.

Tungsten is also used as a conductor in TSV [37, 54, 55]. Tungsten CVD is used for the metal filling process. This may contaminate the silicon, and W-CVD requires a barrier and adhesion layer to be deposited on the SiO_2 surface. TiN is used as a barrier and adhesion layer film to provide good conformality and barrier properties. There will be little plug displacement during repeated thermal cycles to 400°C, since the CTE of tungsten (4.6 ppm/°C) is close to that of bulk Si. In addition, tungsten can decrease the resistance of the interconnections, similar to Cu. Tungsten vias with high yields and excellent reliability have been demonstrated at IBM [54, 56].

Ni results in less contamination in active circuits, has a lower CTE than Cu, and higher electroplating deposition rates than Cu, so plated Ni is a candidates for the conductor material for TSV. Therefore, Ni was tested as the conductive material of the TSV in a newly developed simplified vertical interconnection process [46, 57]. The high-aspect-ratio (AR = 7.5) vias were completely filled without any voids or defects using Ni electroplating [46]. The sulfamate Ni plating bath is a mix of 1 mol/dm³ $Ni(SO_3NH_2)_2$ and 0.65 mol/dm³ H_3BO_3. The Ni electrodeposition was performed at 50°C using a direct current density of 100 mA/cm². The pH value of the plating bath was about 4.0. SEM cross sections of 75 μm deep Ni TSV and Sn-Cu micro-bump are shown in Figure 1.6(d). The first Ni TSVs and Sn-Cu micro-bumps with diameters of 10 μm at a 20 μm pitch were produced without voids.

TSVs that do not affect device reliability are needed even if other materials are chosen for the TSV conductor.

1.4   BONDING TECHNOLOGIES

Due to the need to minimize the wiring length between stacked chips for 3D integration, the demands for new bonding materials and methods are increasing. Several bonding schemes, such as metal bonding [22, 23, 59], SiO_2 direct bonding [10, 60], adhesive bonding [61–63], and hybrid bonding [64, 65] have been proposed to provide reliable vertical interconnections for 3D integration (Table 1.3). This section is an overview of various bonding technologies and processes with their characteristics.

1.4.1   OVERVIEW OF BONDING TECHNOLOGIES

Figure 1.8 shows cross section SEM images of SiO_2 direct bonding, adhesive bonding, metal bonding, and hybrid Cu-adhesive bonding.

SiO_2 direct bonding: SiO_2 direct bonding is used for SOI-based wafer bonding, especially with the face-to-back approach [41]. Before the bonding process, the SOI wafer is polished from the backside until the buried oxide (BOX) layer is exposed. After a hydrophilic layer is formed on the bonding surface, the initially room temperature (RT) bonded wafers are annealed at a high temperature. Covalent bonds are formed in the bonding interfaces through this annealing. The bonding is

$$Si - OH + Si - OH = Si - O - Si + H_2O \qquad (1.6)$$

SiO_2 direct bonding can reduce the effects of misalignments during the bonding process because the initial low-temperature bonding reduces the thermal expansion. However, because of the surface roughness, a high planarization technique is needed so that the roughness is less than 0.4 nm RMS. Both SiO_2 direct bonding and adhesive bonding create simple mechanical bonding. Therefore, via formation and via filling processes are used after the bonding process to electrically connect the stacked layers.

Adhesive bonding: Adhesive bonding is a method to bond wafers with adhesive materials that are then cured. A variety of polymers such as benzocyclobutene (BCB) are used as the adhesive materials. Here is an example of an adhesive bonding process. After preparatory cleaning of the wafer, BCB is spin-coated onto the wafer and soft-baked at 170°C in a flowing N_2 atmosphere. Then pairs of wafers are aligned and bonded at 250°C in a vacuum. The advantage of this approach is that the adhesive can handle particles, surface roughness at the wafer interfaces, and the stress due to the bonding process. However, good procedures for applying the adhesive and curing it are critical for consistent results, and the temperature stability of the resulting bonds is limited compared to other bonding methods.

Hybrid metal-adhesive bonding: For hybrid metal-adhesive bonding, BCB, polyimide, or other polymers are used as an adhesive material [64, 66]. Compared to SiO_2 direct bonding or Cu-Cu bonding, this bonding approach

TABLE 1.3

Bonding Schemes for 3D Integration

Bonding Type	Metal Bonding			SiO₂ Direct Bonding	Adhesive Bonding	Hybrid Bonding (metal and adhesive)
	C4 Bonding (SAC, CuSn, etc.)	IMC Bonding	Cu-Cu Direct Bonding			
Bonding temperature	~ 260°C	~ 260°C	~ 400°C	RT	BCB: <250°C	~300°C
Connectivity	Mechanical and electrical	Mechanical and electrical	Mechanical and electrical	Mechanical	Mechanical	Mechanical and electrical
Advantages	Self-aligning at bonding Good wetting behavior Low bonding pressure	Good thermal stability	Low resistance High thermal conductivity	Low bonding temperature High alignment accuracy @ low temperature	Accommodate particles and surface roughness	Mechanical and electrical connection with adhesive in one processing step
Challenges	Large space and gap Temperature stability	IMC thickness control Reliability	High bonding pressure Flatness control Void-free bonding	Flatness control Void-free bonding	Storage life Limited temperature stability Processing integrity	High bonding pressure Processing integrity Flatness particle

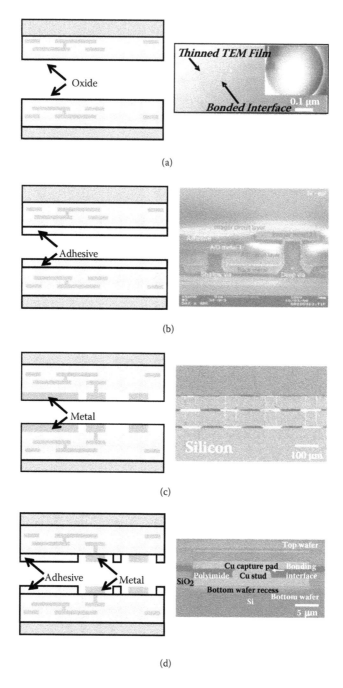

FIGURE 1.8 Examples of bumps fabricated with different technologies: cross section images of (a) oxide-oxide bonded interface [41], (b) adhesive bonding [63], (c) metal bonding [59], and (d) hybrid Cu-adhesive joint [65]. (From Sakuma, *J. Inst. Electrical Eng. Jpn.*, 131(1), 19–25 (2011) (in Japanese). Copyright © 2011 IEEJ.)

has an advantage for bonding conformality due to the adhesive reflow during the bonding. However, high-quality wafer-level surface preparation such as chemical mechanical polishing (CMP), plasma and/or wet cleans is required for the bonding interface, and there are the potential problems with the adhesives, such as moisture absorption or thermal stress due to CTE mismatch.

Metal bonding: Metal bonding including controlled collapse chip connection (C4), low-volume lead-free solder, and direct Cu-to-Cu are the main candidate approaches for the micro-bumps to form electrical interconnections for die-to-die and die-to-wafer 3D integration. Cu-to-Cu bonding has advantages, since Cu has high thermal conductance and low electrical resistance [67–70]. However, Cu-to-Cu bonding requires high temperatures in the range of 350 to 400°C for a Cu diffusion reaction. High pressure and high smoothness are also required. The C4 process has been used in manufacturing for some years [71], but C4 solder interconnects require larger spacing between the balls and larger joint gaps than can be used with low-volume lead-free solder interconnects (IMC bonding). Figure 1.9 compares a C4 solder ball joint with an intermetallic compound (IMC) bonding interconnection. The IMC bonding interconnections are better than C4 interconnections for 3D integration because they have better heat dissipation and allow for smaller design rules in the silicon. In addition, IMC bonding has good thermal stability and good electromigration performance. Therefore, this section focuses on micro-bump technology using IMC bonding. Intermetallic systems and candidate solder materials for IMC bonding are also discussed in the following subsections.

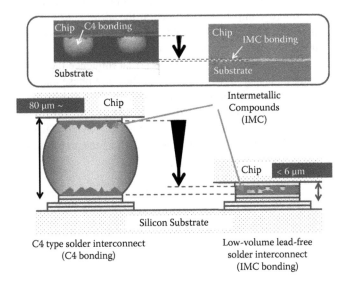

FIGURE 1.9 Comparison between C4 solder interconnect and low-volume lead-free solder interconnect. (From Sakuma et al., *IBM J. Res. Dev.*, 52(6), 611–622 (2008). Copyright © 2008 *IBM Journal of Research and Development*.)

1.4.2   IMC BONDING

The benefits of IMC bonding interconnects include (1) increased vertical heat transfer within a 3D die stack, (2) extension to fine-pitch interconnection design rules, and (3) a temperature hierarchy for low-temperature bonding, but also supporting subsequent process steps with less remelting than C4 interconnects. The temperature hierarchy supports the creation of tested and known-good die stacks without the risk of die stack interconnections melting again during reflow for module-level assembly or surface-mount assembly onto boards. Once created, the IMC bonds have good thermal stability. Low-volume lead-free solder interconnects such as Sn or In are formed and can be joined at relatively low temperature and form an intermetallic phase with a melting temperature much higher than the low bonding temperature [59]. For example, the melting point of In is 156°C, and the melting point of the resulting Cu-In intermetallic compounds is expected to be over 400°C, which is higher than the standard solder reflow temperature (260°C) used in the later bonding steps. This is a desirable feature for 3D die stacks because the high thermal stability supports repetition of the same bonding process steps for more die stack layers or for other subsequent die stack or module-level assemblies. Another advantage of the low-temperature bonding process is the ability to overcome problems, such as wafer or die roughness or warpage during bonding.

1.4.2.1   Intermetallic System

When the solder becomes very thin for the fine-pitch low-volume interconnections of 3D chip integration, the IMC has an important role in controlling the mechanical and electrical integrity of the joints. The joints between the micro-bumps and pads are formed with an interfacial reaction that produces IMC through liquid-solid reactions. During the micro-bump bonding process, the alloy reacts to form complex IMCs depending on the structure and combination of the materials of the solder and pads. The literature describes five factors—atomic size, electronegativity, valence electrons, atomic number, and cohesive energy—that influence the formation of the intermetallic compounds [72]. The properties of the solid phase and the kinetics of the reactions depend primarily on the thermal behavior. The values of these parameters can be estimated from the phase diagrams. At a constant temperature, the growth model as controlled by volume diffusion indicates that the growth dynamics of the interface IMC layer follows a square root law for time, expressed as

$$h = (Dt)^{1/2} \tag{1.7}$$

where D is the diffusion coefficient and t is the aging time. For nonisothermal annealing, the characteristic diffusion length is given by (see [73]):

$$h = \left(\int D(T(t)) dt \right)^{1/2} \tag{1.8}$$

The diffusion coefficient depends on the temperature and is given by an Arrhenius expression,

$$D(T) = D_0 \exp\left(-\frac{Q_0}{RT} \right) \tag{1.9}$$

where D_0 is the diffusion constant, Q_0 is the activation energy for diffusion, R is the gas constant, and T is the absolute temperature. The thickness of the grown IMC can be predicted for each aging condition by using a physical model. The growth of the interface IMC is initially linearly proportional to the square root of the aging time, and a growth model corresponding to Equations (1.7) and (1.8) is valid, but the growth levels off with longer aging times.

By using low-volume lead-free solder (less than 6 µm high), the joints form IMCs in the interfaces of the solder and pads. The IMC system and test vehicle structures for IMC evaluation are summarized in Section 1.4.3.

1.4.2.2   Solder Materials for IMC Bonding

Materials composed of Sn-Pb (tin-lead) have been widely used for soldering package-to-board or chip-to-substrate interconnections. However, the use of lead-based solders is increasingly restricted because of environmental concerns and related legislation [74]. Indium (In) and Sn have low melting temperatures, below 230°C. In research, Cu/Sn, Cu/Ni/In, and Cu/In have been evaluated as lead-free solder interconnection candidates for 3D chip stacks.

1.4.3   CHARACTERISTICS OF IMC BONDING

The mechanical properties of the micro-bumps were evaluated by shear and impact shock testing after the flip chip bonding of the test vehicles. SEM imaging was also used to study the morphology of the IMC layers in the solder joints of the micro-bumps. Detailed descriptions of these experimental results are presented in the next subsections.

1.4.3.1   Test Vehicle for Mechanical Evaluation of IMC Bonding

The silicon chips were joined to a silicon substrate under a forming gas using a precision flip chip bonder, without flux. The bump interconnects have a circular pad shape of thickness less than 10 µm instead of the typical solder ball shape. To control the bump heights precisely and to achieve uniform heights for reliable low-volume solder bonding, an evaporation method is used for the micro-bumps. Bumps with diameters of 100 µm on 200 µm pitch (4-on-8) were fabricated on the silicon chip. Cu/Sn and Cu/Ni/In bumps formed uniformly by an evaporation process were about 3 µm/3 µm and 3 µm/0.5 µm/2 µm across and in height, respectively (Figure 1.10(a)). The pad structures in the silicon substrates contained an outer layer of Ni, beneath a thin layer of immersion Au.

1.4.3.2   Shear Strength

To determine the shear strength per bump, mechanical shear tests were performed on the bonded samples. All of the shear tests were carried out at room temperature. The relationship between shear strength per bump and bonding temperature is shown in Figure 1.10(b). Cu-In intermetallic compounds form above 156°C. The shear strength of Cu/In was found to be lower than that of Cu/Sn. In addition, there were optimal bonding temperatures to get maximum shear strength per bump for Cu/Ni/In, but the shear strength of Cu/Sn joints was relatively unchanged by the bonding temperature. A possible explanation for the increase in shear strength per bump with temperature is

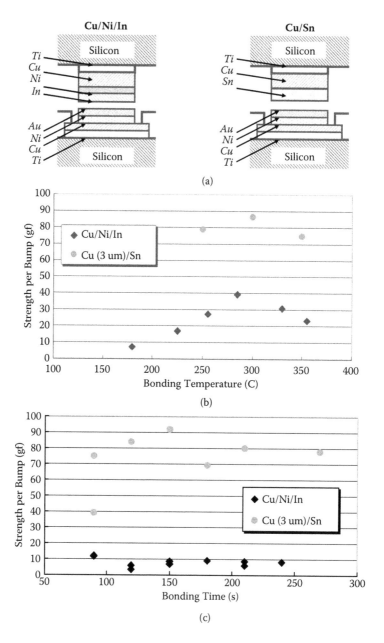

FIGURE 1.10 Mechanical characteristics of IMC bonding: (a) Schematic illustration of IMC bonding joint. (b) Shear strength per bump as a function of bonding temperature. Cu/Ni/In bonding conditions of 2.9 Kgf/cm² and 150 s, and Cu/Sn with bonding conditions of 4.1 Kgf/cm² and 90 s. (c) Shear strength per bump as a function of bonding time. Cu/Ni/In with bonding conditions of 180°C and 2.9 Kgf/cm², and Cu/Sn with bonding conditions of 350°C and 4.1 Kgf/cm². (From Sakuma et al., *IBM J. Res. Dev.*, 52(6), 611–622 (2008). Copyright © 2008 *IBM Journal of Research and Development.*)

that increasing the temperature also increases the reaction rate, so more intermetallic compound is formed during the chosen bonding time (150 s in this case). For Cu/Ni/In and Cu/Sn, when above the optimal bonding temperature (for a chosen bonding time) a thicker nonductile intermetallic layer is probably formed, and this can lead to interconnections that are less resistant to thermal shock. The Cu/Ni/In interconnect metallurgy is a special case, since the IMC is not formed over all of the pad's surface area because of the Ni layer, as will be explained in Section 1.4.3.4. Figure 1.10(c) shows the relationship between shear strength per bump and bonding time. The shear strength of Cu/Ni/In is shown to have lower bonding time than that of Cu/Sn. However there is a little change in the shear strength with the bonding time. The mechanism determining the optimal bonding time is still unclear, since for all bonded samples the shear strength remained fairly constant. This seems to exclude the explanation that 90 seconds is enough time to bond sufficiently after heating samples, and the shear strength of each sample differs only slightly in the range of 90 to 300 seconds bonding time.

1.4.3.3   Shock Test Reliability

The reliabilities of various chips were studied using simulated heat sinks attached to the top of each chip with epoxy, with two different weights of 27 and 54 g/cm^2. Various shock loading conditions, such as a peak deacceleration of 100 G for 2 ms duration, 200 G for 1.5 ms, and 340 G for 1.2 ms, were used with the JESD 22-B110 service conditions C, D, and E [77]. The components were subjected to five shock pulses of the peak level. The results of the impact shock tests are shown in Table 1.4. The results showed that the samples of eutectic PbSn passed all of the targets in the impact shock tests. However, the samples of Cu/Sn did not pass the targeted impact shock test objectives with the simulated heat sinks of 27 and 54 g/cm^2. The cross section SEM and EDX characterizations showed that intermetallic compounds were formed uniformly across the areas of the juncture. In the same tests, the samples of Cu/Ni/In passed the targeted impact shock testing objectives for 27 g/cm^2. In addition, the samples of eutectic PbSn passed all of the targets in the impact shock tests. The Cu/Ni/In chips, formed without IMCs throughout the pad's volume, had better resistance to impact shocks than the Cu/Sn samples with brittle intermetallic compounds formed in their interfaces, as will be discussed in the next section.

1.4.3.4   Cross Section SEM and EDX Analysis

The lead-free Cu/Ni/In and Cu/Sn interconnects were cross-sectioned and characterized by SEM and EDX in order to analyze the mechanisms of the interconnect microstructure and IMCs formed during the bonding process. Table 1.5 shows the values of some key properties of the selected solders and IMCs.

As shown in Figure 1.11(a), the Cu/Ni/In interconnection formed an IMC with Au and Cu in very small areas, not over all of the pad's surface. This means that after bonding, these bumps have higher melting points than the prebonded bumps. Cu will diffuse through the Ni, even at room temperature. For Cu-In IMCs, there are about 3 IMCs between 30 and 50 atomic percentages of In in the phase diagram (gamma, eta, and phi). This would be difficult to analyze by SEM and EDX, even if they are thick enough. These data show that if the η-phase Cu_7In_4 and δ-phase Cu_7In_3 IMCs are formed in the bonding area, then the bonds are thermally stable up to approximately 630°C.

TABLE 1.4

Results of the Impact Shock Test for Three Different Interconnect Metallurgies

Test Sample	G g/ms	Simulated Heat Sink		
		No Mass Pass/Fail	27 g/cm² Pass/Fail	54 g/cm² Pass/Fail
Cu/Sn	110/2.19	Pass	Fail	Fail
	168/1.65	Pass	*	*
	346/1.25	Pass	*	*
Cu/Ni/In	109/2.20	Pass	Pass	Fail
	181/1.65	Pass	Pass	*
	321/1.26	Pass	Pass	*
Eutectic PbSn	101/2.21	Pass	Pass	Pass
	179/1.65	Pass	Pass	Pass
	322/1.26	Pass	Pass	**

Source: Sakuma et al., *IBM J. Res. Dev.*, 52(6), 611–622 (2008). Copyright © 2008 *IBM Journal of Research and Development.*

Note: The bonding conditions for Cu/Sn were 350°C, 150 s, and 4.1 Kgf/cm², and the bonding conditions for Cu/Ni/In were 285°C, 150 s, and 2.9 Kgf/cm².

* no test.

** glue failure.

TABLE 1.5

Properties of the Selected Solders and Intermetallic Compounds at Room Temperature

		In	Cu	Sn	Cu_6Sn_5	Cu_3Sn
Thermal expansion	10^{-6}/K	32.1	16.8	25	16.3	19
Thermal conductivity	W/m · K	80	394	70	34.1	70.4
Melting point	°C	156.6	1,084	232	415	670
Resistivity	$\mu\Omega$ · cm	8.0	1.55	11.5	17.5	8.93
Young's modulus	Gpa	10	110	50	85.56	108.3
Poisson's ratio		0.45	0.34	0.35	0.309	0.299
Density	g/cm³	7.3	8.93	7.17	8.28	8.90

From Figure 1.11(b) it is clear there is good wetting and bonding for the Cu/Sn interconnections. The solder has reacted with the ball limiting metallurgy (BLM), and an IMC of Cu, Ni, Au, and Sn was formed throughout the pad's volume. The Cu-Sn IMC is mostly the η-phase Cu_6Sn_5 with some alloy additions of Ni and Au in Figure 1.11(b), since the Au and Ni used are very thin compared to the thickness of the Cu and Sn. If ε-phase Cu_3Sn is formed, then it will only be a very small amount. Sn is detected on the bonding surface, and it appears that the Sn has diffused throughout

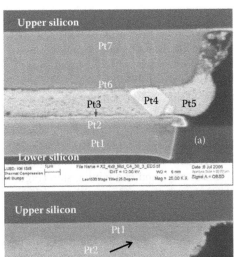

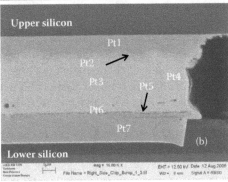

| Pt 1 = Cu |
| Pt 2 = Ni |
| Pt 3 = In, Au, Ni |
| Pt 4 = In, Au, Cu |
| Pt 5 = In |
| Pt 6= Ni |
| Pt 7 = Cu |

| Pt 1 = Cu |
| Pt 2 = Cu,Sn |
| Pt 3 = Ni,Cu,Au,Sn |
| Pt 4 = Ni,Cu,Au,Sn |
| Pt 5 = Ni, P (dark region P rich - depletion of Ni) |
| Pt 6 = NiP |
| Pt 7 = Cu |

FIGURE 1.11 Cross section SEM and EDX analysis for (a) Cu/Ni/In, with bonding conditions of 180°C, 150 s, and 2.9 Kgf/cm². (b) Cu/Sn, with bonding conditions of 350°C, 90 s, and 4.1 Kgf/cm². (From Sakuma et al., *IBM J. Res. Dev.*, 52(6), 611–622 (2008). Copyright © 2008 *IBM Journal of Research and Development.*)

the Cu region of the bump and has formed intermetallic compounds, based on the grain size and activation energy. From the phase diagram, the first intermetallic compound that forms between Sn and Cu is ε-phase Cu_3Sn that corresponds to the phase area at 38 wt% Sn, based on the grain boundary diffusion, a brittle intermetallic that tends to form at the Cu surface [78]. Then there is an equilibrium compound η-phase Cu_6Sn_5 that corresponds to the phase area at 61 wt% Sn based on bulk diffusion. A P concentration is also detected in the bonding areas. This is probably because the electroless Ni-plated layer contains P, and the hypophosphite (H_2PO_2) used as the reducing agent is incorporated into the layer. Electroless Ni plating using nickel sulfate with a hypophosphite reduction has been reported:

$$Ni^{2+} + (H_2PO_2)^- + H_2O \rightarrow Ni^0 + 2H^+ + H(HPO_3)^- \qquad (1.10)$$

It was found that Cu/Ni/In has limited wetting for the IMC to the substrate. However, Cu/Sn IMC was formed throughout the pad's volume and the IMC appears to be effectively wetted. This seems to explain why the shear strength of Cu/Sn is higher than that of Cu/Ni/In. However, the formation of an IMC of brittle Cu_3Sn

resulted in worse results in the shock tests. The Cu/Ni/In chips, formed without IMCs throughout the pad volumes, had better resistance to impact shocks than the Cu/Sn samples with brittle IMCs formed at the interface.

1.4.3.5   Thermal Cycle Testing with IMC Bonding

This section focuses on deep thermal cycle (DTC) testing of the die stack systems with the Cu/Ni/In and Cu/Sn IMC bonding interconnections. There is a significant mismatch in the CTEs of the silicon and organic substrates. The different CTEs of these materials within the modules cause large stresses in the low-volume solder joints [17]. Figure 1.12(a) shows the test vehicles for IMC bonding evaluation. Configuration (a-1) has two silicon substrates and one joining step (Figure 1.12(a)). Configuration (a-1) was made with wiring substrates 725 µm thick and TSV substrates with soldered bumps, with thicknesses of 150 and 70 µm. The TSV was a tungsten annular-type conductor with a diameter of 50 µm. In this case, the CTEs of the two substrates are the same, but the small amount of underfill material in the gaps causes some stress during the thermal cycles. In contrast, configuration (a-2) has an additional 1-2-1 organic substrate (buildup, ABF-GX13; core, MCL-E-679FG(R)), which amplifies the stress and provides resistance for the measurement pads. The organic substrate is 400 µm thick and 23 mm square.

The two interconnection metallurgies of Cu/Ni/In and Cu/Sn were considered the materials for IMC bonding for 3D integration. Each joining with a 100 µm diameter has a copper stud 3 µm high, and the bump pitch was 200 µm. For Cu/Sn, the solder layer is 3 µm of Sn. For Cu/Ni/In there are two layers, with thicknesses of 0.5 µm Ni and 2 µm In.

The DTC test was conducted on the Si/Si samples and the Si/Si-on-organic-substrate samples. The temperature range of the tests was –55 to 125°C. The cycle times were 35 min, and the upper and lower soak times were 15 and 10 min, respectively, for DTC. The electrical contact resistances of the samples were monitored *in situ* using a resistance measurement system. There were 1,000 stress cycles.

The preliminary results showed that with the added organic substrate stress, the Cu/Sn samples had the greatest number of failures, while the Cu/Ni/In had fewer failures. Some failed samples with Cu/Sn joints showed large cracks in the top Si dies (Figure 1.12(b)), and some samples showed cracks between the metal pads and the Cu-Sn IMCs. The strength of the Cu-Sn IMC can become greater than the interface or the Si dies. In such cases, there may be interface failures or cracks in the Si dies. This is reasonable, since the yield strength of the Cu/Sn is comparable (approximately 150 Mpa) to that of a Si die (100 to 150 Mpa).

The preliminary results showed that with the Si/Si samples, some failures occurred among the samples with Cu/Sn joints, but no failures were detected in the Cu/Ni/In joint samples. To further investigate the differences between the Cu/Sn and Cu/Ni/In joints, the average changes of the resistances from the initial values were plotted for the Si/Si samples (Figure 1.12(c)). The values of the resistances at the maximum temperature (125°C) were used for this analysis. The error bars in the plots indicate the standard deviations of the values of ΔR. An increase in resistance for the Cu/Sn joint samples in the DTC test was observed. In contrast, for the systems with the Cu/Ni/In joints, the average change in resistance remained under 1% in the DTC test. The spread of the data was larger for the Cu/Sn samples than for the Cu/Ni/In samples.

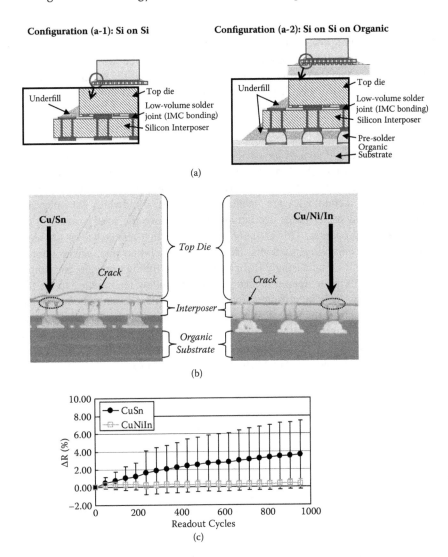

FIGURE 1.12 Thermal cycling test with IMC bonding: (a) Configurations of IMC bonding evaluation test vehicles of (a-1) Si/Si and (a-2) Si/Si on organic substrate. (b) Cross-sectional micrographs of Si/Si on organic substrate samples. (c) The average values of the change in resistance ΔR, from the initial value for the Si/Si systems with Cu/Sn and Cu/Ni/In joints for DTC. (From Sakuma et al., in *Electronic Components and Technology Conference (ECTC)*, 2010, pp. 864–871. Copyright © 2010 IEEE.)

The DTC tests showed that the systems with Cu/Ni/In joints had fewer failures and smaller increases in the electrical resistances of the joints during the tests than the systems with Cu/Sn joints. The In solder joints can act as a stress reliever for these systems, while the strength of the IMCs formed at the Cu/Sn joints can become stronger than the interface strength of the Si dies, and this leads to interfacial failures or cracks in the Si dies. In order to continue improving system-level performance, new insulators such as ultra-low-k (ULK) and air gap materials can be used to reduce dielectric strengths. These materials with lower dielectric strength typically exhibit inferior mechanical properties, such as reduced fracture resistance. Therefore, the low-k material layers are susceptible to damage during thermal cycles, especially if the materials with high yield strengths, such as Cu-Sn IMCs, were used as micro-bumps for 3D integration.

1.4.4 FLUXLESS BONDING

There are increasing demands for smaller packages and finer-pitched interconnections. At the same time, 3D chip integration requires smaller and finer-pitch microbump technologies. The sizes of the vertical interconnections are shrinking, and reliable bonding technologies are needed for high-I/O-bandwidth 3D integration [82]. Conventional solder joints use liquid flux to remove the solid oxide film during the bonding processes. This calls for cleaning the flux residues after the reflow process to avoid reliability problems. However, as the bump pitch shrinks for 3D interconnections, it is increasingly difficult to remove the flux residue. In addition, contractions due to the temperature change from the reflow temperature to the temperature of the flux residue cleaning water cause low-k cracking and chip package interaction (CPI) problems [90]. There are growing demands for fluxless soldering and new surface cleaning methods for fine-pitch 3D interconnects. This subsection describes the results of the comparisons of the treatment methods for Cu (3 μm)/Sn (3 μm) solder bumps and Au pads for fluxless flip chip joints.

1.4.4.1 Ar Plasma Treatment

In flip chip bonding and wire bonding, plasma treatment techniques are generally used to reduce the interface delamination problems caused by organic contamination on the surfaces of the bonding materials [83, 84]. This improves the uniformity and longevity and enhances the bond adhesion strength. However, disadvantages include temperature increases due to heat transfer, poor etching selectivity, and surface damage due to the ion bombardment. In addition, various types of radiation damage caused by the charge buildup can have serious effects on semiconductor devices.

1.4.4.2 VUV Treatment

A vacuum ultraviolet (VUV) surface treatment process is being developed as a cleaning method for the bonding metal surfaces, offering simplicity, low cost, and high productivity. VUV treatment has none of the problems of ion bombardment damage, such as temperature increases, or charge buildup in the plasma treatment. The system uses a xenon excimer (Xe_2*) lamp with a central wavelength of 172 nm for the surface treatment [82]. For the treatment process, the chamber is evacuated after samples are inserted into the chamber and oxygen gas is introduced into the chamber at the desired pressure. The VUV treatment process occurs with chamber pressures below 3.0×10^4 Pa. Below are the reactions at the atomic and molecular levels that are responsible for excimer and excited oxygen generation:

$$Xe_2^* \rightarrow Xe + Xe + h\nu \tag{1.11}$$

$$O_2 + h\nu \rightarrow 2O \tag{1.12}$$

$$O_2 + O \rightarrow O_3 \tag{1.13}$$

$$2O_3 + h\nu \rightarrow 3O_2 \tag{1.14}$$

For the VUV/O_3 treatment, the excimer light interacts with the oxygen in the chamber, and the surface of the sample is affected by the high-density ozone (O_3) and excited oxygen radical O(1D). The O(1D) decomposes any organic matter on the sample surfaces [85, 86]. In addition, the energy of the photons at 172 nm wavelength is 697.5 kJ/mol, and these photons can directly break molecular bonds with bond energies up to 697.5 kJ/mol.

1.4.4.3 Formic Acid Treatment

In cases where a plasma surface treatment method was not used, formic acid vapor has been used as a surface treatment for fluxless flip chip soldering and for metal oxide removal during reflow [87]. When the temperature is above 150°C, the formic acid reacts with solder oxides to form compounds. At temperatures above 200°C, these compounds further decompose into carbon dioxide and hydrogen. Here are the primary reactions of gaseous formic acid with metal oxides [88]:

$$MeO + 2HCOOH \rightarrow Me(COOH)_2 + H_2 \tag{1.15}$$

$$Me(COOH)_2 \rightarrow Me + CO_2 + H_2 \tag{1.16}$$

$$H_2 + MeO \rightarrow Me + H_2O \tag{1.17}$$

In this notation, Me represents an arbitrary metal. To remove an oxidation film on Sn and decompose the compounds formed by the reaction of Sn and formic acid, the bonding samples received a formic acid treatment at 200°C at a chamber pressure of 1,100 mbar [89].

1.4.4.4 Hydrogen Radical Treatment

Dry chemical cleaning including hydrogen radicals (excited hydrogen) is being developed to remove organic compounds, contaminants, and oxide films. This process uses the reducing action of the hydrogen radical. A hot wire method using a tungsten wire filament has been proposed to produce hydrogen radicals, because hydrogen radicals require a high temperature, in the range of 1,500 to 2,000°C [112]. The treatment process takes place at low atmospheric pressure, typically less than 100 mm Torr. In another version, hydrogen radicals are generated by a surface wave microwave (2.45 GHz) plasma in a hydrogen ambient at 1 Torr, and they irradiate the target samples on the stage [90, 113].

1.4.4.5 Comparison of Surface Treatments

Various cleaning conditions with Ar plasma, VUV/O_3, formic acid vapor, and hydrogen radicals were evaluated and compared. Evaporated 3 µm/3 µm thick Cu/Sn and immersion Au plating over electroless Ni plating were used for the bonding micro-bumps and bonding pads. X-ray photoelectron spectroscopy (XPS) was used to study the surface elemental composition of the micro-bumps and pad surfaces before and after the cleaning processes. The photoelectron spectra of C1s and Sn3d were obtained with XPS, and the results showed that the hydrogen radical and VUV/O_3

surface treatments effectively reduced the carbon-based organic contamination on the bonding surface (Figure 1.13(a)). In addition, hydrogen radicals and formic acid treatment can remove oxides effectively, as shown in Figure 1.13(b).

After the surface treatments, the upper and lower silicon pieces were bonded at 260°C without flux by using a bonding tool. The solder connections' shear force was improved by the hydrogen radical surface treatments, as shown in Figure 1.13(c). Hydrogen radical treatments with higher temperatures gave higher average shear strength. In comparing various treatment conditions, the results show that the hydrogen radical treatments and VUV/O_3 treatments result in higher average shear strength

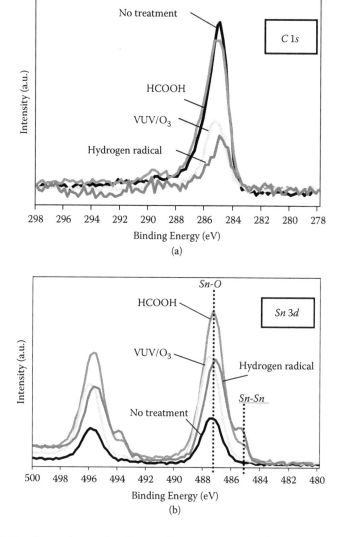

FIGURE 1.13 Comparisons of various surface treatments: XPS narrow-scan spectra of (a) C1s and (b) $Sn3d_{5/2}$ of the flip chip with Cu/Sn bump surfaces, with and without treatments.

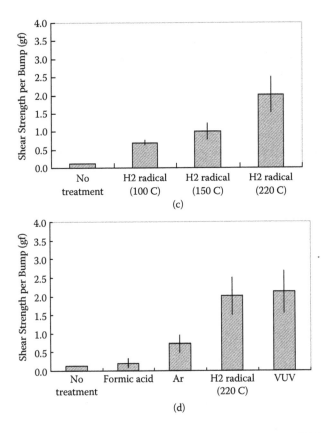

FIGURE 1.13 (*Continued*) Comparisons of various surface treatments: XPS narrow-scan spectra of (c) average shear strength with and without hydrogen radical treatment. (d) Average results of shear tests with various treatments [90].

among the treatments tested, with shear strengths more than 16 times stronger than those of the untreated samples (Figure 1.13(d)) [90]. These preliminary experiments indicate that the hydrogen radical treatments and VUV/O$_3$ treatments are useful for cleaning the bonding interconnections.

1.5 CHIP LEVEL 3D INTEGRATION TECHNOLOGY

For 3D integration with a top-down approach, two or more 2D circuits, correspond-ing to different layers of the 3D devices, are fabricated independently, then aligned and bonded together at the chip level or wafer level for the 3D integration. The TSV for the vertical interconnections between each layer can be fabricated before or after each layer is stacked by using face-to-face or face-to-back integration schemes, as required by the target applications. High-precision alignment and high-quality high-reliability bonding are required when stacking the device layers. Either bulk or SOI-based CMOS should be used for stacking the layers, since the applications can affect the performance requirements, vertical interconnection lengths, and device densities. It is best if heterogeneous chips or wafers can be combined and connected with high-density vertical interconnections. There are several different approaches for 3D integration, including die-to-die, die-to-wafer, and wafer-to-wafer. Schematic illustrations of wafer-to-wafer and die-to-wafer integration processes are shown in Figure 1.14.

Die-to-die and die-to-wafer: Die-to-die and die-to-wafer integration can be done with high-precision flip chip bonding. In general, the die sizes will be different when multiple technologies are used for the assembling. Die-to-die and die-to-wafer integration technologies make it possible to stack multiple known-good dies (KGDs) with different die sizes in layers [91].

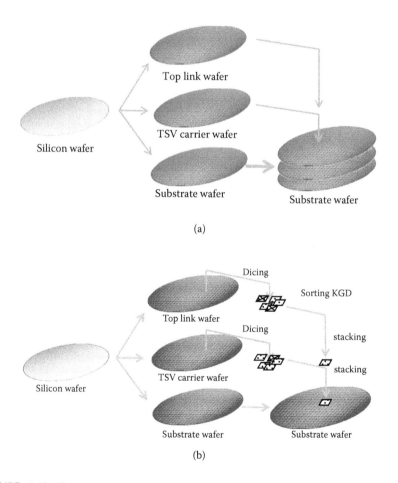

FIGURE 1.14 Schematic illustration of 3D integration process: (a) wafer-to-wafer 3D integration process, (b) die-to-wafer 3D integration process. (From Sakuma et al., in *58th Electronic Components and Technology Conference (ECTC)*, Lake Buena Vista, FL, 2008, pp. 18–23. Copyright © 2008 IEEE.)

However, when dies are bonded as arrays on a wafer, the process has to be repeated as many times as there are laminated dies in the array. If precise alignment accuracy is required, the processing time for stacking each layer increases. Due to the low expected fabrication throughput, die-to-die and die-to-wafer bonding techniques may ultimately not be cost-effective. Die-to-die and Die-to-wafer technologies can be used when high yield or dies of different sizes are needed for the 3D integration.

Wafer-to-wafer: Wafer-to-wafer integration technology may provide the ultimate solution for the highest manufacturing throughput, depending on achieving sufficiently high yields and minimal loss of good die and wafers [92, 93]. The use of wafer-to-wafer integration technology is most suitable with high-yield wafers and small die sizes. This calls for raising the yields to much higher levels than are currently possible. In addition, all of the layers must use wafers of the same diameter and compatible technologies, since all of the layers must be aligned at the wafer level. The materials and geometries will be complicated because any mismatches due to such factors as thermal gradients between the stacked wafers will cause displacements during the bonding processes. The total yield of 3D integration using wafer-to-wafer technology is determined by multiplying the yield of each wafer [94]. Therefore, the compound die yield decreases exponentially as the number of stacked layers increases. This technology could be used as long as the die yield of each wafer is sufficiently high.

1.5.1 DIE YIELD OF STACKING PROCESSES

The total yield of 3D stack applications using wafer-to-wafer technology is most constrained by the wafer with the lowest yield, since the total yield is determined by multiplying the yields of each wafer. This also means the chip yield will decrease exponentially as the number of stacked layers increases. The physically possible number of chips (N_c) produced from a wafer may be given as [95]

$$N_c = \pi \frac{\left[\phi - (1+\theta)\sqrt{A/\theta}\right]^2}{4A} \tag{1.18}$$

where

$$A = x \cdot y \tag{1.19}$$

and

$$\theta = \frac{x}{y} \geq 1 \tag{1.20}$$

In Equations (1.18) to (1.20), x and y are the dimensions of a rectangular chip (mm), θ is the ratio between x and y, ϕ is the wafer diameter (mm), and A is the area of the chip (mm^2). For example, for a 300 mm wafer with $x = 10$ mm and $y = 10$ mm, $N_c = 615$ chips. If defect density remains constant, then the chip yield decreases as the chip size increases. Therefore, small chip size is desirable to raise the yield of a wafer. To think about yield of the 3D integration, it is necessary to consider the yield of the stacking process as well as the yield of the wafer. When combining n untested chips with a chip yield Y_{2D}, the compound yield of the wafer-to-wafer structure will be given by [96]

$$Y = Y_{2D}^n \cdot Y_S^{n-1} \tag{1.21}$$

where Y_S is the yield of the stacking process. For example, combining three dies with wafer-to-wafer integration with a device yield, Y_{2D}, of 85% and a stacking yield, Y_S, of 95%, results in a module yield of only 55%. Lower wafer yields result in exponentially smaller module yields.

In the future, with structure and process optimizations, wafer-to-wafer integration may provide a solution for the highest throughput, but that assumes or targets applications where a high yield is possible. In the near term, die-to-die or die-to-wafer integration may offer high yield, high flexibility, and high performance plus time-to-market advantages. Compared to the die-to-wafer processes recently developed, a new integration technology, called cavity alignment method (CALM), that significantly reduces the complexity of fabrication is being developed. This technology reduces the number of processing steps in fabricating die stacks.

1.5.2 CAVITY ALIGNMENT METHOD (CALM)

Chip level 3D integration offers the ability to stack KGD, which can lead to higher yields without integrated redundancy and flexible combinations of different technologies. Conventional die-to-die and die-to-wafer bonding are performed using the flip chip method, which involves aligning an array of solder interconnects with pads on the substrate, followed by application of heat and pressure. When joining multiple dies, it is necessary to repeat the positioning, heating, and pressure steps once for each of the dies to be joined. Also, when joining stacked dies in an array on a wafer, the process has to be repeated as many times as the number of joined dies arranged in the array.

In this section, die-to-wafer integration using CALM is introduced. The technology is inexpensive with high throughput, and supports a high-precision automatic positioning technique [97, 98]. Figure 1.15(a) shows a schematic illustration of the die-to-die 3D integration process with CALM. The cavity holds a die and can be used as a positioning reference when joining multiple dies. Heat and pressure are applied to join the stacked die in one step, utilizing the cavity as a reference surface to align the entire stack. Therefore, the manufacturing throughput is greatly enhanced and no mechanical device is necessary to provide positioning, so three-dimensional stacked dies can be manufactured at low cost.

The number of dies to be stacked is n, the number of stacked dies placed on the wafer is m, the processing time for alignment is P_a, and the processing time for pressurizing, heating, and cooling is P_h, so the total processing time for conventional die-to-wafer integration is given as

$$P_t = (P_a \times n \times m) + (P_h \times n \times m) \tag{1.22}$$

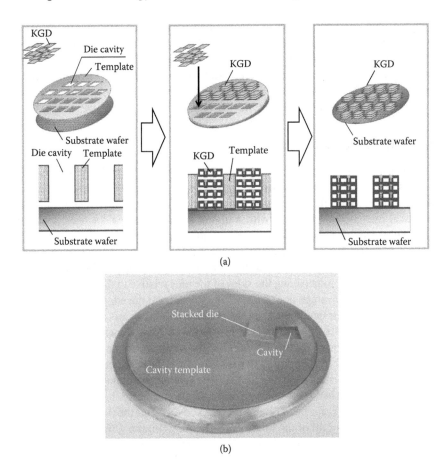

(a)

(b)

FIGURE 1.15 Cavity alignment method (CALM): (a) Schematic illustration of the die-to-wafer 3D integration process with CALM. (b) Photo image of the die cavity prototype. (From Sakuma et al., in *58th Electronic Components and Technology Conference (ECTC)*, Lake Buena Vista, FL, 2008, pp. 18–23. Copyright © 2008 IEEE.)

In contrast, the total processing time for die-to-wafer integration using the CALM approach is described by

$$P_t = (P_c \times n \times m) + P_h \tag{1.23}$$

where P_c is the processing time to place the dies inside cavities and $P_c \ll P_a$. The die-to-wafer 3D integration manufacturing throughput can be significantly improved by using die cavity, since the heat and pressure for the entire die are applied in one step and the stacked chips are created on the wafer. All of the processes can be done at the same time, and equipment with alignment functions for each die is no longer needed. Also, the bonding thermal history of each chip in the wafer can be ignored, even when multiple dies are stacked.

To explore the feasibility of the die-to-wafer integration process, a die cavity prototype was built. Figure 1.15(b) shows a photograph of a prepared prototype cavity, which has only one hole. This hole is controlled in size in comparison to the actual die size to control the alignment, assembly, and release. It is important that the template material is compatible with the processing, the CTE, and the alignment tolerance control of the die and die stacks. Many options may be considered for the template material, such as metals, ceramics, and silicon. The use of silicon or silicon derivatives can help control the CTE during processing.

1.5.3 ALIGNMENT ACCURACY

We evaluated the alignment accuracy of the die cavity technology [111]. The in-plane positions of the solder joints relative to the die edges before and after stacking were measured to evaluate the alignment accuracy of the cavity alignment technology. An optical microscope was used with a digital position readout for these measurements. The accuracy of the measurements with this microscope is estimated as $\pm 1\ \mu m$.

The size of the chips and the cavity was measured first. The average width of the silicon dies was 2,968 μm and the width of the cavity was 2,992 μm. Prior to stacking, the positions x_i of the joints along the sides of the dies relative to the edges were measured. We found the average position x_0 of the joints from the edge for each group of dies to be stacked. The deviation Δx of the position x_i from x_0 was calculated for each side of the dies (Figure 1.16(a)), and the histogram of Δx is shown in Figure 1.16(a). The standard deviation of Δx was 1.1 mm, and the result shows a narrow distribution.

The chip stacks were cut parallel to the sides of the dies. The cross sections of the chip stacks were polished after the dicing. The relative positions x_i of the joints in the plane of the polished cross sections were measured. From these measurements, the average position x_0 for each row of the stacked dies and the deviations Δx of the positions x_i from x_0 for all of the joints were obtained (Figure 1.16(a)). The deviation Δx for a row of joints is obtained from the average of the values for all of the joints in the same row. The rotational deviations are negligible, since the standard deviations of the Δx within each row were less than 1 μm. A histogram of the deviation Δx after stacking is plotted in Figure 1.16(b). The standard deviation of the values of Δx before stacking was 1.1 μm, and the value after stacking was 2.0 μm.

The results show that precision in chip alignment of less than 5 μm was achieved, even after three-layer chip stacking using the CALM.

1.5.4 TEST VEHICLE: DESIGN AND FEATURES

The primary electrical test vehicle for the chip level 3D integration was designed as a TSV carrier and a substrate. To measure the electrical resistance and yield of the TSV and lead-free solder interconnects, the test vehicle consisted primarily of daisy chains with links alternating between bottom and top dies. The diameter of each via was 50 μm. The diameter of each bump was 100 μm and the pitch was 200 μm. An annular W via is defined, since this has been previously shown to give low-resistance, high-yield TSVs, which are easily integrated into a standard CMOS copper BEOL process flow [99]. Wiring links were deposited with 2 μm thickness of an electroplated Cu layer followed by an electroless Ni layer and 100 μm of an immersion Au layer deposited to form chains in the substrate.

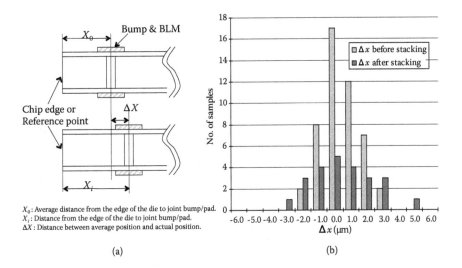

X_0: Average distance from the edge of the die to joint bump/pad.
X_i: Distance from the edge of the die to joint bump/pad.
ΔX: Distance between average position and actual position.

(a)   (b)

FIGURE 1.16 Alignment accuracy of die cavity technology: (a) Schematic drawings of the cross sections of the dies showing the deviation Δx for the joints from their average value x_0. (b) Histogram of the deviation from the average position of the bumps before and after 3-layer stacking. For the histogram of the measurements before stacking, both edges of the side were measured for each x and y direction, whereas only one measurement per direction of the side is taken after stacking. The positions x_0 are the averages among the stacked samples. (From Sakuma et al., in *64th Electronic Components and Technology Conference (ECTC)*, Lake Buena Vista, FL, 2014, pp. 647–654. Copyright © 2014 IEEE.)

Detailed descriptions of the process flow of this technology appear elsewhere [37, 59, 99] and are only summarized here. First, the pattern of annular vias with diameters of 50 μm is etched into the silicon substrate with ICP RIE, and insulation is formed on the sidewalls of the vias using thermal oxidation. As an electrically conducting material, tungsten is deposited by CVD followed by standard planarization with CMP. Single damascene Cu pads with a 200 μm pitch are formed on top of the TSV. After electroless Ni and immersion Au deposition to form the top surface metallurgy (TSM) of the receiving pads, a glass handling wafer is attached to the surface for mechanical support during wafer grinding. The wafers are thinned to the thickness of 70 μm by mechanical grinding and CMP, and the bottoms of the tungsten-filled vias are exposed. Following an insulation process on the backside using PECVD, a final CMP step exposes the via metal. The bottom surface metallurgy (BSM) and Cu/Ni/In lead-free solder metal are defined on the back surface by evaporation through an aligned metal mask. The lead-free solder interconnects are less than 6 μm in height.

1.5.5 RESULTS OF CHIP STACKING USING CALM

The alignment between the die and a substrate is performed using CALM. The dies to be stacked are placed inside the controlled cavities. All dies are bonded simultaneously in the bonding process. The lamination and bonding process is performed in controlled equipment with a fixture that provides uniform heating. The lamination cycle utilizes a controlled temperature, pressure, and ambient atmosphere to form the bonds. Through this batch fabrication, all of the dies on one wafer can be joined in parallel and high-throughput alignment and bonding can be done.

By using a lamination tool and a cavity for die bonding, a chip level 3D integration process was demonstrated between TSV carriers and a silicon substrate.

Figure 1.17(a) shows a SEM image of a six-layer die stacked on a supporting silicon substrate. The vertical die stack on the substrate appears to be precisely aligned along the line of edges. The pads on the surface of the supporting substrate are used for contact with the backside bump interconnects to the chip stack and also for probe testing. Figure 1.17(b) shows a cross-sectional SEM image of a six-layer chip stack on a supporting silicon substrate. This figure shows that the annular via and lead-free solder bumps are connected vertically. Figure 1.17(c) shows an optical image of three-layer dies stacked at multiple sites on a supporting silicon substrate. The dies at multiple sites were stacked simultaneously without any bonding failures

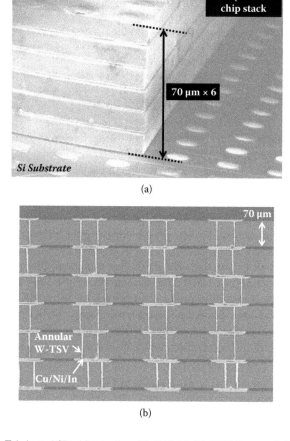

(a)

(b)

FIGURE 1.17   Fabricated 3D chip stacks with CALM: (a) SEM image of six-layer stack on a Si substrate, (b) cross-sectional SEM image of a chip stack.

(c)   (d)

FIGURE 1.17 (*Continued*)   Fabricated 3D chip stacks with CALM: (c) optical image of three-layer dies stacked at multiple sites on a substrate with CALM, (d) cross section SEM image of thermal evaluation 3D chip. (From Sakuma et al., *IBM J. Res. Dev.*, 52(6), 611–622 (2008). Copyright © 2008 *IBM Journal of Research and Development.*)

due to the die cavity technology. These early results demonstrated the initial feasibility for stacking dies with 200 µm pitch interconnections.

Die cavity technology was also used to make 3D chip stacks to evaluate the thermal resistances of 3D chips for different structures and layouts of the TSV and micro-bumps [100]. The 3D chip stack samples with thermal monitoring sensors using diode, Cu TSV, and SnAg micro-bumps were fabricated with the die cavity assembly technology (Figure 1.17(d)). The diode on the bottom chip is electrically connected with the pads on the top chip through the TSV and micro-bumps. There is no difference in the diode characteristics in the top and the bottom layer after chip stacking. These results show that the stacking process, along with the distortion caused by the stacking, does not have any influence on the characteristics of the diodes after joining.

1.5.6 ELECTRICAL TESTS

Electrical tests were performed to measure the yield and average electrical resistance of each tungsten TSV and Cu/Ni/In lead-free solder interconnect using the IBM test vehicle's variable-length chains. Figure 1.18 shows the measured DC resistance of the link chains in the 3D chip-stacking test vehicles. A total of six different locations were measured for one-layer, three-layer, and six-layer dies stacked on supporting silicon substrates. They were fabricated by using the chip level 3D integration process. The total resistance of the link chains, R_{total}, is given as

$$R_{total} = 2n(R_{via} + R_{bump}) + R_{wiring} \qquad (1.24)$$

where n is the number of stacked layers, R_{via} is the resistance of the TSV, R_{bump} is the resistance of a micro-bump, and R_{wiring} is the resistance of the wiring on the substrate. As shown in Figure 1.18(b), the measured total resistance of the link chains is indicated on the right axis, and the resistance of a single TSV plus lead-free solder interconnects on the left axis. The total resistance shown in Figure 1.18(b) includes the annular tungsten TSV, the Cu/Ni/In bumps, and the Cu wiring links that are the dominant components in the total resistance. These data show the average resistance of a single tungsten TSV and Cu/Ni/In solder interconnect was approximately 21 mΩ, which is an acceptably low value.

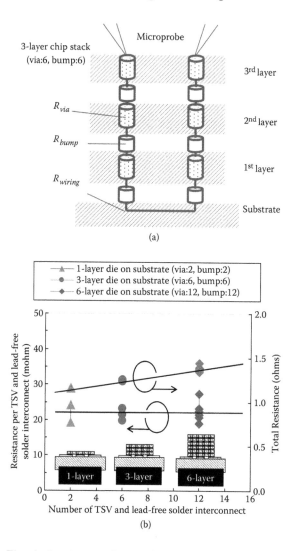

FIGURE 1.18 Electrical resistance characteristics of tungsten TSV and Cu/Ni/In bump: (a) Electrical resistance measurement for 3-layer chip stack. (b) Electrical resistance of one-layer, three-layer, and six-layer dies stacked on supporting silicon substrates. (From Sakuma et al., *IBM J. Res. Dev.*, 52(6), 611–622 (2008). Copyright © 2008 *IBM Journal of Research and Development*.)

1.6 UNDERFILL FOR CHIP LEVEL 3D INTEGRATION

Conventional underfill to fill in the gaps under a flip chip package is done with an underfill flow via capillary action. This process may have limitations, especially as the chips become larger in size and as the standoff gap is reduced. In addition, it is difficult to fill thin gaps (less than 10 μm, which are required for 3D integration) without voids and high filler content, which reduces the adhesive coefficient of thermal expansion and can increase thermal conductivity.

1.6.1 OVERVIEW OF UNDERFILL PROCESS

The alternatives to conventional capillary underfill are wafer-level underfill [101, 102] and no-flow underfill [103, 104].

The wafer-level underfill approach has the potential of being a fast and low-cost operation since the underfill is precoated on the wafer and B-staged before it is diced into chips for the joining processes. A number of wafer-level underfill approaches have already been described using various materials and processing steps. The challenge for this technology is to optimize the materials and processing flow to achieve high yield and reliable bonds. This technology approach must overcome challenges such as trapped air bubbles, poor wetting to the solder bumps, or poor alignment because the underfill material may cover up the alignment marks.

No-flow underfill provides flux to the solder during bonding and eliminates the extra processing steps, such as flux residue cleaning required for a capillary process. This is rapid process suitable for mass production, since the device can be kept at an elevated temperature until the underfill is cured to prevent low-k cracking. One of the challenges is to avoidable inclusions of filler particles in the solder joints, which can affect the reliability of these connections.

Vacuum underfill can be considered an extension of the capillary underfill process where the filled adhesive flow is enhanced using gas pressure [105]. In this approach, an underfill is dispensed around the 3D stacked chip or flip chip under reduced pressure in a vacuum chamber. When the vacuum is released, the pressure within the chips is now lower than the external atmospheric pressure. The pressure difference assists the insulating underfill material in penetrating into the narrow gaps between the stacked chips. To date, there have been few experimental studies on the application of underfill with vacuum assistance for 3D chip stacks. Matsumoto et al. proposed an adhesive injection method for 3D LSI [106, 107]. However, their process cannot form fillet around stacked chips. In addition, their process requires a wall around the target area to create a pressure difference, and the wall must be designed as part of the layout design. The vacuum underfill technology does not need any wall, which makes this process more suitable for manufacturing.

1.6.2 VACUUM UNDERFILL PROCESS

Figure 1.19 shows a comparison of standard capillary underfill deposition with the vacuum underfill process for 3D chip stacks. For the vacuum underfill process, the stacked chips were placed in the vacuum chamber before dispensing the underfill material [105]. The stage temperature of the vacuum underfill tool was controlled, and this temperature depends on the properties of the underfill material. The vacuum chamber includes a dispensing device for the underfill material. After placing the sample on the stage in the vacuum chamber, the chamber is evacuated. Then the underfill material is dispensed around each stacked chip on the substrate, and the vacuum is released. When the vacuum is filled with air at normal atmospheric pressure, the underfill, which is dispensed all around each stacked chip, is injected by air pressure into the narrow gaps between each chip.

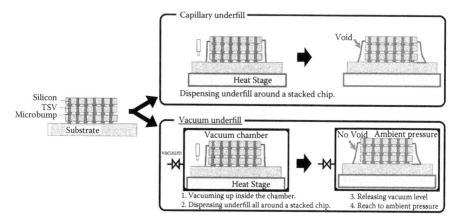

FIGURE 1.19 Comparison of standard capillary underfill deposition with vacuum underfill process for 3D chip stack.

1.6.3 Results of Vacuum Underfill for 3D Chip Stack

Vacuum underfill technology was evaluated for 3D chip stacks. A primary electrical test vehicle for the 3D chip stack explained in Section 1.5.4 was used. The test vehicle consisted of wired daisy chains. The three layers were vertically stacked by using CALM [97]. The bonding was done with a controlled bonding temperature, time, pressure, and ambience. With this technique, all of the chips are stacked in one step. Each layer is electrically connected by tungsten TSVs and Cu/Sn bumps. The thickness of each stacked chip is approximately 70 μm, and the bumps are 100 μm in diameter with a pitch of 200 μm and a height of 6 μm. Electrical tests were done to measure the resistance and yield of each vertical interconnection of the three-layer chip stacks with underfill. The results showed the average resistance of a single tungsten TSV and Cu/Sn micro-bump was around 75 mΩ, and no failures occurred in any of the chains. To investigate the thermal reliability of the stacked chip, one-layer TSV stacks on silicon substrates with and without underfill were subjected to the following reliability test conditions [77]:

1. JEDEC level 3 moisture preconditioning:
   a. 125°C bake for 24 h
   b. 30°C at 60% relative humidity for 192 h
   c. Three times at 260°C peak reflow
2. Deep thermal cycle: From –55 to 125°C at a rate of two cycles per hour

Figure 1.20 is a summary of the thermal reliability results up to 1,500 cycles for the stacked chips. There was minimal change in the average resistances between the chip stacks with and without underfill.

Figure 1.21(a) shows a SEM image of a three-layer chip stack on a silicon substrate before vacuum underfill process. As shown in the pictures, the chips are precisely aligned along their edges. Figure 1.21(b) shows an optical microscope image of the chip

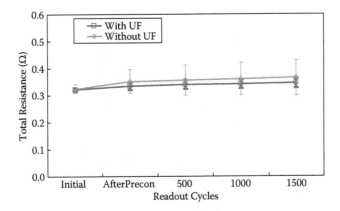

FIGURE 1.20 DTC results for the chip stack with and without underfill. Four-point resistance of the paired tungsten TSVs and Cu/Sn bumps (including the wiring in a one-layer chip stack).

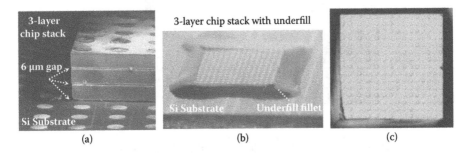

FIGURE 1.21 A 3D chip stack sample before and after vacuum underfill process: (a) three-layer chip stack without underfill, (b) with underfill. Fillet formed around the chip stack; (c) SAM image after vacuum underfill process.

stack after vacuum underfill process. An underfill that includes filler particles (with an average particle size of 0.3 μm and a maximum particle size of 1 μm) was used. The filler content is 55% by weight. The underfill should completely fill the gaps between each stacked thin chip and its substrate. The objectives for this underfill study included good adhesion, no voids, and the formation of fillets around the stacked chips. As shown in the picture, this process forms a well-shaped fillet around the chip stack. Figure 1.21(c) shows a scanning acoustic microscope (SAM) image of the three-layer chip stack with underfill after 1,000 DTCs. There are no delaminations or large voids in the underfill of the chip stack. The thermal reliability test results showed that the resistances of the chip stack with tungsten TSVs and Cu/Sn bumps with and without underfill were acceptable.

1.7 STACKING OF LARGE DIE WITH 22 NM CMOS DEVICES

3D chip level integration technology has been demonstrated with large Si die in 22 nm CMOS technology [111]. The size of the die was more than 600 mm^2. Figure 1.22(a) shows cross-section SEM image of a 3D integration with a 22 nm CMOS die stack

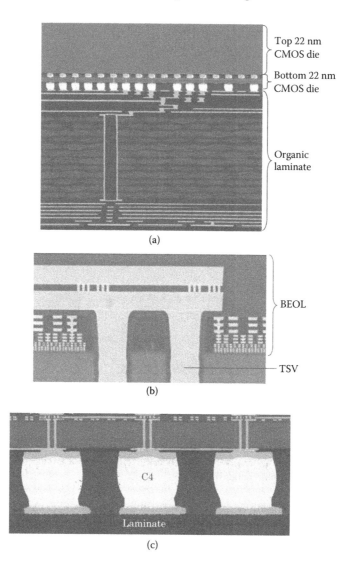

FIGURE 1.22 Cross section SEM image of (a) face-to-face die stack on organic laminate fabricated with IBM's 22 nm process. The size of the die was more than 600 mm².
(b) Integrated TSV and BEOL structure (22 nm), (c) C4s assembled using chip level 3D
integration process. (From Sakuma et al., in *64th Electronic Components and Technology
Conference (ECTC)*, Lake Buena Vista, FL, 2014, pp. 647–654. Copyright © 2014 IEEE.)

on a laminate. Dies were assembled on a laminate with 7 layers of build-up circuitry
on each side of the core. Warpage of thin TSV die is not ideal for conventional
furnace reflow. Laminate substrates experience additional warping during heating
because each laminate layer can have different CTE. The influence of thin die warpage on 3D integration becomes more significant as die size increases and component
thickness decreases. The challenges related to maintaining thin die and laminate

co-planarity were overcome by using a new assembly technology. Figure 1.22(b) shows a cross-section SEM of an integrated TSV and BEOL structure. Cu-TSV is introduced in the Back-End-of-Line (BEOL). The vias are completed with bottom-up Cu electroplating to fill the TSVs. Void-free TSVs were formed and no delamination of the BEOL dielectric stack was observed. Additional BEOL levels are fabricated after TSV processing and subsequently planarized by CMP. Two types of bumps were used for the interconnection: low-solder volume with Cu pillars for the top die to the bottom die, and high-solder volume for the bottom die to the laminate. The top die consisting of Cu and SnAg was joined to Ni/Au pads on the bottom die side, while the bottom die side had the traditional SnAg bump with terminal Ni joined to pre-solder on the laminate side. Figure 1.22(c) shows a cross-section SEM image of the assembled C4 joints. The gap between bottom die and laminate is stable and no joint failure was observed.

Table 1.6 shows the stress conditions that were run to evaluate the 3D modules and test results. The stresses were carried out to the durations specified by the JEDEC standard [77]. The sample size and number of lots that were sampled are also shown in Table 1.6. The tests during readouts included comprehensive continuity and leakage testing of various macros that were designed to evaluate the integrity of the upper-bump-TSV-lower-bump connection, the laminate wiring to those structures, and the integrity of the BEOL wiring levels in both chips in a 3D format.

The thermal reliability of thermal interface material level-1 (TIM1) and the die-to-die interfaces is also evaluated by using the same test vehicle. The test vehicle was constructed with vertically-aligned heaters and 25 thermal sensors (resistance thermal device, RTD) embedded in the BEOL layers of each die. A thermocouple was attached to the center outside surface of the lid as a package reference temperature sensor. The traditional interface from the top of the chip (stack) to the package lid through the TIM1, often labeled Theta J-C or Rint, is monitored using the 25 top-chip sensors referenced to the external lid-attached thermocouple. Unique to the stacked-chip structure is a new thermal interface in the interconnection level between the die as visible in Figure 1.22(a). The thermal resistance parameter associated with this interface is labeled as Resistance die-to-die (Rd-d) and is calculated using the temperature difference measured between each of the vertically-aligned temperature sensors embedded in each BEOL stack. As shown in Figure 1.23, both Rint and Rd-d were extremely

TABLE 1.6

Reliability Stress Conditions for Chip Level 3D Modules and Test Results

Cell	Stress	Condition	JEDEC Spec	Requirements	Qty (Lots)	Result
A	TC-G	−40/125°C	A104	850 cy/0 Fail	13 (1)	Pass
B	THB	85°C /85% RH/3.6V	A101	1000 hr/0 Fail	30 (2)	Pass
C	HTS	150°C	A103	1000 hr/0 Fail	13 (2)	Pass

Source: Sakuma et al. in *64th Electronic Components and Technology Conference (ECTC)*, Lake Buena Vista, FL, 2014, pp. 647-654. Copyright © 2014 IEEE.)

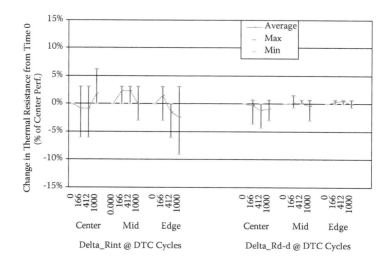

FIGURE 1.23   Thermal Reliability of TIM1 (Rint) and die-to-die (Rd-d) interfaces through 1000cy of DTC (−55/+125°C) Accelerated Stressing. (From Sakuma et al., in *64th Electronic Components and Technology Conference (ECTC)*, Lake Buena Vista, FL, 2014, pp. 647–654. Copyright © 2014 IEEE.)

stable throughout the duration of stressing with a maximum thermal degradation of less than 4°C per 1000W of power dissipation observed at the end-of-stress readout.

1.8 SUMMARY

This chapter has described the development trends and key technologies of chip level 3D integration. 3D integration is emerging as an approach to achieve high bandwidth, high performance, high functionality, and reduced complexity in the interconnections of electronic circuits. Through experiments, the foundations of the key technologies for 3D integration have been devised, such as TSV, IMC bonding, and fluxless bonding for effective chip bonding, advanced chip level 3D integration processes with high yield and high throughput based on CALM, and vacuum underfilling technology. Reliability results showed that the TSV and IMC bonding have excellent electrical connectivity and thermal performance, allowing high density and flexible interconnections for constructing high-performance 3D systems. In addition, large thin Si die stacking with 22 nm CMOS devices has been demonstrated. The integrity of the bump-TSV connection, the BEOL wiring levels, and the laminate wiring has been proven.

ACKNOWLEDGMENTS

The author thanks the following people for their contributions to this work: J.U. Knickerbocker, P.S. Andry, C.K. Tsang, R.R. Horton, S.K. Kang, B. Dang, S.L. Wright, C.S. Patel, B.C. Webb, J. Maria, E.J. Sprogis, R.J. Polastre, R. Sirdeshmukh, D. Dimilia, and M. Farinelli, IBM T.J. Watson Research Center; S. Skordas, J. Zitz,

E. Perfecto, W. Guthrie, H. Liu, and G. Advocate, IBM Semiconductor Research & Development Center; K. Sueoka, S. Kohara, K. Matsumoto, K. Toriyama, A. Horibe, H. Noma, Y. Orii, and K. Kawase, IBM Research–Tokyo; Y. Oyama, T. Aoki, H. Nishiwaki, and S. Jacobs, IBM Japan; L. Guerin, and R. Langlois, IBM Canada; Prof. S. Shoji, Prof. J. Mizuno, Prof. H. Kawarada, Prof. T. Tanii, Prof. T. Watanabe, H. Ono, N. Nagai, N. Unami, and M. Nimura, Waseda University; Prof. M. Koyanagi and Prof. K.W. Lee, Tohoku University. The author also acknowledges the support of the IBM Central Scientific Services. In addition, the author thanks management for support, including K. Sikka, M. Angyal, D. Berger, S. Iyer, B. Sundlof, J. Trewhella, and T.C. Chen, IBM Research.

PUBLICATION ACKNOWLEDGMENTS

The original sources of the material in the chapter follow.
This chapter is a revised version of:

Katsuyuki Sakuma / Kris Iniewski, *Nano-Semiconductors, Devices and Technology* (Chapter 7: Development of 3D chip integration technology), pp. 173–221, ISBN9781439848357, CRC press, 2011. [Reproduced by permission of Taylor and Francis Group, LLC, a division of Informa plc.]
Katsuyuki Sakuma, Three-dimensional chip integration technology for high density packaging, PhD thesis, Waseda University, Tokyo, Japan, 2009.

Section 1.7 is a revised version of:

K. Sakuma, S. Skordas, J. Zitz, E. Perfecto, W. Guthrie, L. Guerin, R. Langlois, H. Liu, K. Ramachandran, W. Lin, K. Winstel, S. Kohara, K. Sueoka, M. Angyal, T. Graves-Abe, D. Berger, J. Knickerbocker, and S. Iyer, Bonding technologies for chip level and wafer level 3D integration, In *64th Electronic Components and Technology Conference (ECTC)*, 2014, pp. 647–654.

REFERENCES

1. K. Sakuma, Three-dimensional chip integration technology for high density packaging, PhD thesis, Waseda University, Tokyo, Japan, 2009.
2. L. Lev and P. Chao, *Down to the wire, requirements for nanometer design implementation*, Cadence White Paper, 2002.
3. W. Haensch, Why should we do 3D integration? In *45th Design Automation Conference (DAC)*, 2008, pp. 674–675.
4. P. Brofman, IBM's packaging Technology Roadmap and the "collaboratory" approach to advanced packaging development, In *International Conference on Electronics Packaging (ICEP)*, 2009, pp. 1–6.
5. J.A. Kahle, M.N. Day, H.P. Hofstee, C.R. Johns, T.R. Maeurer, and D. Shippy, Introduction to the cell multiprocessor, *IBM J. Res. Dev.*, 49(4/5), 589–604 (2005).
6. K. Bernstein, P. Andry, J. Cann, P. Emma, D. Greenberg, W. Haensch, M. Ignatowski, S. Koester, J. Magerlein, R. Puri, and A. Young, Interconnects in the third dimension: design challenges for 3D ICs, In *Proceedings of Design Automation Conference (DAC)*, June 2007.

7. S. Borkar, Thousand core chips: a technology perspective, In *Design Automation Conference (DAC)*, 2007, pp. 746–749.
8. V. Suntharalingam, R. Berger, J.A. Burns, C.K. Chen, C.L. Keast, J.M. Knecht, R.D. Lambert, K.L. Newcomb, D.M. O'Mara, D.D. Rathman, D.C. Shaver, A.M. Soares, C.N. Stevenson, B.M. Tyrrell, K. Warner, B.D. Wheeler, D.-R.W. Yost, and D.J. Young, Megapixel CMOS image sensor fabricated in three-dimensional integrated circuit technology, In *International Solid-State Circuits Conference (ISSCC)*, San Francisco, CA, 2005, pp. 356–357.
9. H. Kurino, K.W. Lee, T. Nakamura, K. Sakuma, K.T. Park, N. Miyakawa, H. Shimazutsu, K.Y. Kim, K. Inamura, and M. Koyanagi, Intelligent image sensor chip with three dimensional structure, In *International Electron Devices Meeting (IEDM)*, 1999, pp. 879–882.
10. A.W. Topol, B.K. Furman, K.W. Guarini, L. Shi, G.M. Cohen, and G.F. Walker, Enabling technologies for wafer-level bonding of 3D MEMS and integrated circuit structures, In *Proceedings of the 54th Electronic Components and Technology Conference (ECTC)*, 2004, pp. 931–938.
11. M. Ieong, Three dimensional CMOS device and integrated circuits, In *Proceedings of the IEEE Custom Integrated Circuits Conference*, 2003, pp. 207–213.
12. S.F. Al-sarawi, A review of 3-D packaging technology, *IEEE Trans. Comp. Pkg. Manuf. Technol. B*, 21, 1 (1998).
13. S. Das, A.P. Chandrakasan, and R. Reif, Calibration of Rent's rule models for three-dimensional integrated circuits, *IEEE Trans. VLSI Syst.*, 12(4), 359–366 (2004).
14. B.S. Landman and R.L. Russo, On a pin versus block relationship for partitions of logic graphs, *IEEE Trans. Comput.*, C-20, 1469–1479 (1971).
15. A. Rahman and R. Reif, System level performance evaluation of three-dimensional integrated circuits, *IEEE Trans. VLSI* (Special Issue), 8(6), 671–678 (2000).
16. S.M. Sri-Jayantha, G. McVicker, K. Bernstein, and J.U. Knickerbocker, Thermomechanical modeling of 3D electronic packages, *IBM J. Res. Dev.*, 52(6), 2008.
17. K. Sakuma, K. Sueoka, Y. Orii, et al., IMC bonding for 3D interconnection, In *Electronic Components and Technology Conference (ECTC)*, 2010, pp. 864–871.
18. S. Das, A. Chandrakasan, and R. Reif, Design tools for 3-D integrated circuits, In *Proceedings of the 2003 Conference on Asia South Pacific Design Automation*, January 21–24, 2003.
19. S.S. Sapatnekar, Physical design automation challenges for 3D ICs, In *Proceedings of the International Conference on Integrated Circuit Design and Technology*, 2006, p. 172.
20. S. Mysore et al., Introspective 3D chips, In *Proceedings of International Conference on Architectural Support for Programming Languages and Operating Systems (ASPLOS XII)*, 2006, pp. 264–273.
21. C. Ryu, B. Banijamali, I. Mohammad, C. Wade, V. Oganesian, and K. Endo, µPILR embedded package technology for mobile applications, *International Wafer-Level Packaging Conference*, 2008.
22. M. Koyanagi, H. Kurino, K.W. Lee, K. Sakuma, N. Miyakawa, and H. Itani, Future system-on-silicon LSI chips, *IEEE MICRO*, 18(4), 17–22 (1998).
23. A. Klumpp, R. Merkel, R. Wieland, and P. Ramm, Chip-to-wafer stacking technology for 3D system integration, In *Proceedings of the 53rd Electronic Components and Technology Conference (ECTC)*, New Orleans, LA, 2003, pp. 1080–1083.
24. J.U. Knickerbocker, P.S. Andry, L.P. Buchwalter, E.G. Colgan, J. Cotte, H. Gan, R.R. Horton, S.M. Sri-Jayantha, J.H. Magerlein, G. Manzer, G. McVicker, C.S. Patel, R.J. Polastre, E. Sprogis, C.K. Tsang, B.C. Webb, and S.L. Wright, System-on-package (SOP) technology, characterization and applications, In *Proceedings of the 56th Electronic Components and Technology Conference (ECTC)*, 2006, pp. 415–421.

25. R. Zingg, J.A. Friedrich, G.W. Neudeck, and B. Hofflinger, Three-dimensional stacked MOS transistors by localized silicon epitaxial overgrowth, *IEEE Trans. Electron Devices*, 37, 1452–1461 (1990).
26. G.W. Neudeck, S. Pae, J.P. Denton, and T. Sue, Multiple layers of silicon-on-insulator for nanostructure devices, *J. Vacuum Sci. Technol. B*, 17(3), 994–998 (1999).
27. S. Pae, T. Su, J.P. Denton, and G.W. Neudeck, Multiple layers of silicon-on-insulator islands fabrication by selective epitaxial growth, IEEE Electron Device Lett., 20(5), 194–196 (1999).
28. S. Kawamura, N. Sasaki, T. Iwai, M. Nakano, and M. Takagi, Three-dimensional CMOS IC's fabricated by using beam recrystallization, *IEEE Electron Device Lett.*, EDL-4(10), 366–368 (1983).
29. K. Sugahara, T. Nishimura, S. Kusunoki, Y. Akasaka, and H. Nakata, SOI/SOI/bulk-Si triple-level structure for three dimensional devices, *IEEE Electron Device Lett.*, 7(3), 193–195 (1986).
30. T. Kunio, K. Oyama, Y. Hayashi, and M. Morimoto, Three dimensional ICs, having four stacked active device layers, *IEDM Tech. Dig.*, 837–840 (1989).
31. A. Kohno, T. Sameshima, N. Sano, M. Sekiya, and M. Hara, High performance poly-Si TFTs fabricated using pulsed laser annealing and remote plasma CVD with low temperature processing, *IEEE Trans. Electron Device*, 42, 251–257 (1995).
32. K.W. Guarini, A.W. Topol, M. Ieong, R. Yu, L. Shi, M.R. Newport, D.J. Frank, D.V. Singh, G.M. Cohen, S.V. Nitta, D.C. Boyd, P.A. O'Neil, S.L. Tempest, H.B. Pogge, S. Purushothaman, and W.E. Haensch, Electrical integrity of state-of-the-art 0.13 μm SOI CMOS devices and circuits transferred for three-dimensional (3D) integrated circuit (IC) fabrication, In *International Electron Devices Meeting (IEDM)*, 2002, pp. 943–945.
33. J.U. Knickerbocker, P.S. Andry, B. Dang, R. Horton, M. Interrante, C.S. Patel, R. Polastre, K. Sakuma, R. Sirdeshmukh, E. Sprogis, S. Sri-Jayantha, C. Tsang, B. Webb, and S.L. Wright, Three-dimensional silicon integration, *IBM J. Res. Dev.* 52(6), 553–570 (2008).
34. K. Sakuma, J. Mizuno, and S. Shoji, 3D chip-stacking technology, In *2009 Lithography Workshop*, 2009, p. 40.
35. M. Feil, C. Adler, D. Hemmetzberger, M. Konig, and K. Bock, The challenge of ultra thin chip assembly, In *Electronic Components and Technology Conference*, 2004, pp. 35–40.
36. S.J. Koester, A.M. Young, R.R. Yu, S. Purushothaman, K.-N. Chen, D.C. La Tulipe Jr., N. Rana, L. Shi, M.R. Wordeman, and E.J. Sprogis, Wafer-level 3D integration technology, *IBM J. Res. Dev.*, 52(6), 583–597 (2008).
37. P.S. Andry, C.K. Tsang, B.C. Webb, E.J. Sprogis, S.L. Wrigth, B. Dang, and D.G. Manzer, Fabrication and characterization of robust through-silicon vias for silicon-carrier applications, *IBM J. Res. Dev.*, 52(6), 2008.
38. T. Fukushima, Y. Yamada, H. Kikuchi, T. Tanaka, and M. Koyanagi, Self-assembly process for chip-to-wafer three-dimensional integration, In *Electronic Components and Technology Conference (ECTC)*, 2007, pp. 836–841.
39. H. Quinones, A. Babiarz, L. Fang, and Y. Nakamura, Encapsulation technology for 3D stacked packages, In *International Conference on Electronics Packaging (ICEP)*, 2002.
40. P. Ramm, A. Klumpp, R. Merkel, J. Weber, R. Wieland, A. Ostmann, and J. Wolf, 3D system integration technologies, In *Materials Research Society Symposium Proceedings*, Boston, MA, 2003.
41. A.W. Topol, D.C. La Tulipe Jr., L. Shi, D.J. Frank, K. Bernstein, S.E. Steen, A. Kumar, G.U. Singco, A.M. Young, K.W. Guarini, and M. Ieong, Three-dimensional integrated circuits, *IBM J. Res. Dev.*, 50(4), 491–504 (2006).
42. K. Takahashi, Y. Taguchi, M. Tomisaka, H. Yonemura, M. Hoshino, M. Ueno, Y. Nemoto, Y. Yamaji, H. Terao, M. Umemoto, K. Kameyama, A. Suzuki, Y. Okayama, T. Yonezawa, and K. Kondo, Process integration of 3D chip stack with vertical interconnection, In *Proceedings of the 54th Electronic Components and Technology Conference (ECTC)*, 2004, pp. 601–609.

43. J. Hopkins, H. Ashraf, J.K. Bhardwaj, A.M. Hynes, I. Johnston, and J.N. Shepherd, The benefits of process parameter ramping during the plasma etching of high aspect ratio silicon structures, In *Materials Research Society (MRS) Symposium Proceedings (MRS)*, 1998.
44. C. Gormley, K. Yallup, W. Nevin, J. Bharadwaj, H. Ashraf, P. Huggett, and S. Blackstone, State of the art deep silicon anisotropic etching on SOI bonded substrates for dielectric isolation and MEMS applications, In *ECS 5th International Wafer Bonding Symposium*, 1999, pp. 350–361.
45. Bosch Gmbh R.B. 1994 U.S. Patent 4855017, U.S. Patent 4784720, and Germany Patent 4241 045C1 (1994).
46. K. Sakuma, N. Nagai, J. Mizuno, and S. Shoji, Simplified 20-um pitch vertical interconnection process for 3D chip stacking, *IEEJ Trans. Electrical Electronic Eng.*, 4, 339–344 (2009).
47. P. Marchal, B. Bougard, E. Beyne, et al., 3-D technology assessment: path-finding the technology/design sweet-spot, *Proc. IEEE*, 97(1), 2009.
48. B. Vandevelde, C. Okoro, M. Gonzalez, B. Swinnen, and E. Beyne, Thermo-mechanics of 3D-wafer level and 3D stacked IC packaging technologies, In *9th International Conference on Thermal, Mechanical and Multiphysics Simulation and Experiments in Micro-Electronics and Micro-Systems (EuroSimE)*, 2008, pp. 1–7.
49. J.U. Knickerbocker, P.S. Andry, L.P. Buchwalter, A. Deutsch, R.R. Horton, K.A. Jenkins, et al., Development of next-generation system-on-package (SOP) technology based on silicon carriers with fine-pitch chip interconnection, *IBM J. Res. Dev.*, 49(4/5), 2005.
50. N. Tanaka, T. Sato, Y. Yamaji, T. Morifuji, M. Umemoto, and K. Takahashi, Mechanical effects of copper through-vias in a 3D die-stacked module, In *Proceedings of the Electronic Components and Technology Conference (ECTC)*, 2002, pp. 473–479.
51. M. Tomisaka, H. Yonemura, M. Hoshino, et al., Electroplating Cu filling for through-vias for three-dimensional chip stacking, In *Proceedings of the 52nd Electron Components and Technology Conference*, San Diego, CA, 2002.
52. J.-S. Yang, K. Athikulwongse, Y.-J. Lee, et al., TSV. Stress aware timing analysis with applications to 3D-IC layout optimization, In *ACM Design Automation Conference*, 2010.
53. K.W. Lee, T. Nakamura, K. Sakuma, K.T. Park, H. Shimazutsu, N. Miyakawa, K.Y. Kim, H. Kurino, and M. Koyanagi, Development of three-dimensional integration technology for highly parallel image-processing chip, *Jpn. J. Appl. Phys.*, 39, 2473–2477 (2000).
54. C.K. Tsang, P.S. Andry, E.J. Sprogis, C.S. Patel, B.C. Webb, D.G. Manzer, and J.U. Knickerbocker, CMOS-compatible silicon through-vias for 3D process integration, In *Materials Research Society Symposium Proceedings*, Boston, MA, 2006.
55. Y. Igarashi, T. Morooka, Y. Yamada, T. Nakamura, K.W. Lee, K.T. Park, H. Kurino, and M. Koyanagi, Filling of tungsten into deep trench using time-modulation CVD method, In *Proceedings of International Conference on Solid State Devices and Materials (SSDM)*, 2001, pp. 34–35.
56. K. Sakuma, P.S. Andry, B. Dang, C.K. Tsang, C. Patel, S.L. Wright, B. Webb, J. Maria, E. Sprogis, S.K. Kang, R. Polastre, R. Horton, and J.U. Knickerbocker, 3D chip-stacking technology with through-silicon vias and low-volume lead-free interconnections, *IBM J. Res. Dev.*, 52(6), 611–622 (2008).
57. K. Sakuma, H. Ono, N. Nagai, J. Mizuno, and S. Shoji, A new fine-pitch vertical interconnection process for through silicon vias and microbumps, In *Asia-Pacific Conference on Transducers and Micro-Nano Technology (APCOT)*, Taiwan, Tainan, June 2008.
58. K. Sakuma, Development trend of three-dimensional (3D) integration technology, *J. Inst. Electrical Eng. Jpn.*, 131(1), 19–25 (2011) (in Japanese).
59. K. Sakuma, P.S. Andry, B. Dang, J. Maria, C. Tsang, C. Patel, S.L. Wright, B. Webb, E. Sprogis, S.K. Kang, R. Polastre, R. Horton, and J. Knickerbocker, 3D chip stacking technology with low-volume lead-free interconnections, In *Proceedings of the 57th Electronic Components and Technology Conference (ECTC)*, 2007, pp. 627–632.

60. K. Warner, J. Burns, C. Keast, R. Kunz, D. Lennon, A. Loomis, W. Mowers, and D. Yost, Low-temperature oxide-bonded three-dimensional integrated circuits, In *Proceedings of IEEE International SOI Conference*, 2002, pp. 123–125.

61. P. Ramm, A. Klumpp, R. Merkel, J. Weber, R. Wieland, A. Ostmann, and J. Wolf, 3D system integration technologies, *Mater. Res. Soc. Symp. Proc.*, 766 (2003).

62. R.J. Gutmann, J.-Q. Lu, S. Devarajan, A.Y. Zeng, and K. Rose, Wafer-level three-dimensional monolithic integration for heterogeneous silicon ICs, In *Proceedings of the IEEE Topical Meeting on Silicon Monolithic Integrated Circuits in RF Systems*, Atlanta, GA, 2004, pp. 45–48.

63. J. Burns, L. McIlrath, C. Keast, et al., Three-dimensional integrated circuits for low-power, high bandwidth systems on a chip, In *International Solid-State Circuits Conference (ISSCC)*, 2001, pp. 268–269.

64. R.J. Gutmann, J.J. McMahon, S. Rao, F. Niklaus, and J.-Q. Lu, Wafer-level via-first 3D integration with hybrid-bonding of Cu/BCB redistribution layers, In *Proceedings of the International Wafer Level Packaging Congress (IWLPC)*, 2005, pp. 122–127.

65. F. Liu, R.R. Yu, A.M. Young, J.P. Doyle, X. Wang, L. Shi, K.-N. Chen, X. Li, D.A. Dipaola, D. Brown, C.T. Ryan, J.A. Hagan, K.H. Wong, M. Lu, X. Gu, N.R. Klymko, E.D. Perfecto, A.G. Merryman, K.A. Kelly, S. Purushothaman, S.J. Koester, R. Wisnieff, and W. Haensch, A 300-mm wafer-level three-dimensional integration scheme using tungsten through-silicon via and hybrid Cu-adhesive bonding, In *International Electron Devices Meeting (IEDM)*, 2008.

66. M. Nimura, K. Sakuma, S. Shoji, and J. Mizuno, Solder/adhesive bonding using simple planarization technique for 3D integration, In *61st Electronic Components and Technology Conference (ECTC)*, 2011, pp. 1147–1152.

67. B. Swinnen, W. Ruythooren, P. De Moor, L. Bogaerts, L. Carbonell, K. De Munck, B. Eyckens, S. Stoukatch, D. Sabuncuoglu Tezcan, Z. Tökei, J. Vaes, J. Van Aelst, and E. Beyne, 3D integration by Cu-Cu thermo-compression bonding of extremely thinned bulk-Si die containing 10 μm pitch through-Si vias, In *Proceedings of the International Electron Devices Meeting (IEDM)*, San Francisco, CA, 2006.

68. K.-N. Chen, S.H. Lee, P.S. Andry, C.K. Tsang, A.W. Topol, Y.-M. Lin, J.-Q. Lu, A.M. Young, M. Ieong, and W. Haensch, Structure, design and process control for Cu bonded interconnects in 3D integrated circuits, In *Proceedings of the International Electron Devices Meeting*, San Francisco, CA, 2006.

69. P. Morrow, M.J. Kobrinsky, S. Ramanathan, C.-M. Partk, M. Harmes, V. Ramachandrarao, H.-M. Park, G. Kloster, S. List, and S. Kim, Wafer-level 3D interconnects via Cu bonding, In *Proceedings of the UC Berkeley Extension Advanced Metallization Conference*, Berkeley, CA, 2004, pp. 125–130.

70. R. Reif, C.S. Tan, A. Fan, K.N. Chen, S. Das, and N. Checka, 3-D interconnects using Cu wafer bonding: technology and applications, In *Advanced Metallization Conference (AMC)*, San Diego, 2002, pp. 37–45.

71. S.L. Wright, R. Polastre, H. Gan, L.P. Buchwalter, R. Horton, P.S. Andry, E. Sprogis, C.S. Patel, C.K. Tsang, J.U. Knickerbocker, J.R. Lloyd, A. Sharma, and M.S. Sri-Jayantha, Characterization of micro-bump C4 interconnects for Si-carrier SOP applications, In *Proceedings of the 56th Electronic Components and Technology Conference (ECTC)*, San Diego, CA, 2006, pp. 633–640.

72. J.H. Westbrook and R.L. Fleischer, (Eds.) *Intermetallic Compounds 1*, 1994, pp. 91–125, 227–275, John Wiley & Sons, New York.

73. P. Shewmon, *Diffusion in solids*, Minerals, Metals & Materials Society, 1989, p. 37.

74. V. Eveloy, S. Ganesan, Y. Fukuda, and M.G. Ji Wu Pecht, Are you ready for lead-free electronics? *IEEE Trans. Comp. Pkg. Manuf. Technol.*, 28(4), 884–894 (2005).

75. L.S. Goldmann, Geometric optimization of controlled collapse interconnections, *IBM J. Res. Dev.*, 251–265 (1969).

76. B. Hochlowski, et al., Low-cost wafer bumping using C4NP, In *Future Fab International*, 18(19), 2005.
77. JEDEC Solid State Technology Association, Electronic Industry Association, http://www.jedec.org/Home/
78. D. Frear, et al., Microstructural observations and mechanical behavior of Pb-Sn solder on copper plates, *Mater. Res. Soc. Symp. Proc.*, 72, 181–186 (1986).
79. Rika nenpyo (chronological scientific tables), Maruzen, Tokyo, 2000.
80. R.J. Fields and S.R. Low, Physical and mechanical properties of intermetallic compounds commonly found in solder joints, Research Publication, National Institute of Standards and Technology, Metallurgy Division, February 2002.
81. L.H. Van Vlack, *Elements of Materials Science and Engineering* (6th ed.), Addison-Wesley Publishing Company, Reading, MA, 1989.
82. K. Sakuma, J. Mizuno, N. Nagai, N. Unami, and S. Shoji, Effects of vacuum ultraviolet surface treatment on the bonding interconnections for flip chip and 3-D integration, *IEEE Trans. Electronics Pkg. Manuf.*, 33(3), 2010.
83. G. Nicolussi and E. Beck, Plasma chemical cleaning of chip carrier in a downstream hollow cathode discharge, In *IEEE Advanced Semiconductor Manufacturing Conference*, 2002, pp. 172–176.
84. J.-M. Koo, J.-B. Lee, Y.J. Moon, W.-C. Moon, and S.-B. Jung, Atmospheric pressure plasma cleaning of gold flip chip bump for ultrasonic flip chip bonding, *J. Phys. Conf. Ser.*, 100, 012034 (2008).
85. K. Sakuma, N. Nagai, J. Mizuno, and S. Shoji, Vacuum ultraviolet (VUV) surface treatment process for flip chip and 3-D interconnections, In *59th Electronic Components and Technology Conference (ECTC)*, 2009, pp. 641–647.
86. N. Unami, K. Sakuma, J. Mizuno, and S. Shoji, Effects of excimer irradiation treatment on thermocompression Au-Au bonding, *Jpn. J. Appl. Phys.*, 49, 6, 2010.
87. R.A. Nordin et al., High performance optical data link array technology, In *Proceedings of 43rd Electronic Components and Technology Conference*, 1993.
88. W. Lin and Y.C. Lee, Study of fluxless soldering using formic acid vapor, *IEEE Trans. Adv. Pkg.*, 22(4), 1999.
89. K. Sakuma, N. Unami, S. Shoji, and J. Mizuno, Fluxless bonding using vacuum ultraviolet and formic acid for 3D interconnects, In *Materials Research Society (MRS)*, 2010, 1249-F09-04.
90. K. Sakuma, K. Toriyama, H. Noma, K. Sueoka, N. Unami, S. Shoji, J. Mizuno, and Y. Orii, Fluxless bonding for fine-pitch and low-volume solder 3-D interconnection, In *61st Electronic Components and Technology Conference (ECTC)*, 2011, pp. 7–13.
91. C. Scheiring, H. Kostner, P. Lindner, and S. Pargfrieder, Advanced-chip-to-wafer technology: enabling technology for volume production, In *International Microelectronics and Packaging Society (IMAPS)*, Long Beach, CA, November 14–18, 2004.
92. R.J. Gutmann, J.-Q. Lu, S. Devarajan, A.Y. Zeng, and K. Rose, Wafer-level three-dimensional monolithic integration for heterogeneous silicon ICs, In *Silicon Monolithic Integrated Circuits in RF Systems*, 2004, pp. 45–48.
93. R. Chatterjee, M. Fayolle, P. Leduc, S. Pozder, B. Jones, E.B. Acosta, B. Charlet, T. Enot, M. Heitzmann, M. Zussy, A. Roman, O. Louveau, S. Maitrejean, D. Louis, N. Kernevez, N. Sillon, G. Passemard, V. Pol, V. Mathew, S. Garcia, T. Sparks, and Z. Huang, Three dimensional chip stacking using a wafer-to-wafer integration, In *International Interconnect Technology Conference*, 2007, pp. 81–83.
94. E. Beyne, 3D system integration technologies, In *Symposium on VLSI Technology*, 2006, pp. 1–9.
95. J.H. Lau, *Low Cost Flip Chip Technologies*, McGraw-Hill.
96. E. Beyene, Solving technical and economical barriers to the adoption of through-Si-via 3D integration, In *Electronics Packaging Technology Conference (EPTC)*, 2008, pp. 29–34.

97. K. Sakuma, P.S. Andry, C.K. Tsang, K. Sueoka, Y. Oyama, C. Patel, B. Dang, S.L. Wright, B. Webb, E. Sprogis, R. Polastre, R. Horton, and J.U. Knickerbocker, Characterization of stacked die using die-to-wafer integration for high yield and throughput, In *58th Electronic Components and Technology Conference (ECTC)*, Lake Buena Vista, FL, 2008, pp. 18–23.

98. K. Sakuma, P.S. Andry, C.K. Tsang, Y. Oyama, C.S. Patel, K. Sueoka, E.J. Sprogis, and J.U. Knickerbocker, Die-to-wafer 3D integration technology for high yield and throughput, In *Materials Research Society (MRS)*, Boston, MA, December 2008.

99. P.S. Andry, C.K. Tsang, E. Sprogis, C. Patel, S.L. Wright, B.C. Webb, L.P. Buchwalter, D. Manzer, R. Horton, R. Polastre, and J.U. Knickerbocker, A CMOS-compatible process for fabricating electrical through-vias in silicon, In *Proceedings of the 56th Electronic Components and Technology Conference*, San Diego, CA, 2006, pp. 831–837.

100. K. Matsumoto, S. Ibaraki, K. Sakuma, K. Sueoka, H. Kikuchi, Y. Orii, and F. Yamada, Thermal characterization of a three-dimensional (3D) chip stack based on experiments and simulation, In *Mate*, 2011 (in Japanese).

101. L. Nguyen, H. Nguyen, A. Negasi, Q. Tong, and S.H. Hong, In *SEMI/IEEE IEMT*, 2002, pp. 53–62.

102. C. Feger, N. LaBianca, M. Gaynes, S. Steen, Z. Liu, R. Peddi, and M. Francis, In *Proceedings of ECTC*, 2009, pp. 1502–1505.

103. R. Agarwal, W. Zhang, P. Limaye, and W. Ruythooren, In *Proceedings of ECTC*, 2009, pp. 345–349.

104. P.L. Tu, Y.C. Chan, and K.C. Hung, Reliability of microBGA assembly using no-flow underfill, *Microelectronics Reliability*, 41(11), 1867–1875 (2001).

105. K. Sakuma, S. Kohara, K. Sueoka, Y. Orii, M. Kawakami, K. Asai, Y. Hirayama, and J.U. Knickerbocker, Development of vacuum underfill technology for 3-D chip stack, *J. Micromech. Microeng*, 21, 035024 (2011).

106. T. Matsumoto, M. Satoh, K. Sakuma, H. Kurino, N. Miyakawa, H. Itani, and M. Koyanagi, New three-dimensional wafer bonding technology using the adhesive injection method, *Jpn. J. Appl. Phys.*, 1(3B), 1217–1221 (1998).

107. K. Sakuma, K.W. Lee, T. Nakamura, H. Kurino, and M. Koyanagi, A new wafer-scale chip-on-chip (W-COC) packaging technology using adhesive injection method, In *Conference on Solid State Devices and Materials (SSDM)*, 1998, pp. 286–287.

108. A. Telikepalli, Power-Performance Inflection at 90 nm Process Node- FPGAs in Focus, Chip Design, http://www.chipdesignmag.com/display.php?articleId=261.

109. R. Bashir, S. Venkatesan, and G.W. Neudeck, A polysilicon contacted subcollector BJT for a three-dimensional BiCMOS process, *IEEE Electron Device Letters,* 13, No. 8, 392–395 (1992).

110. K.-W. Lee, J. C. Bae, T. Fukushima, T. Tanaka, and M. Koyanagi, Evaluation of Cu diffusion characteristics at backside surface of thinned wafer for reliable three-dimensional circuits, *Semicond. Sci. Technol.*, vol. 26, no. 2, p. 025007, Feb. 2011.

111. K. Sakuma, S. Skordas, J. Zitz, E. Perfecto, W. Guthrie, L. Guerin, R. Langlois, H Liu, K. Ramachandran, W. Lin, K. Winstel, S. Kohara, K. Sueoka, M. Angyal, T. Graves-Abe, D. Berger, J. Knickerbocker, and S. Iyer, Bonding technologies for chip level and wafer level 3D integration, In *64th Electronic Components and Technology Conference* (ECTC), 2014, pp. 647–654.

112. K. Abe and A. Izumi, A novel copper interconnection cleaning by atomic hydrogen using diluted hydrogen gas, *J. Solid State Phenomena*, 145–146 (2009) pp. 389–392.

113. T. Hagihara, T. Takeuchi, et al., Fluxless flip-chip bonding process using hydrogen radical, *Electronics Packaging Technology Conference*, pp. 595–600, 2008.

114. K. W. Lee, H. Wang, J.C. Bae, M. Murugesan, Y. Sutou, T. Fukushima, T. Tanaka, J. Koike, and M. Koyanagi, Barrier properties of CVD Mn oxide layer to Cu diffusion for 3-D TSV, IEEE *Electron Device Letters*, Vol. 35, No.1, 2014.

2 Wafer-Level Three-Dimensional ICs for Advanced CMOS Integration

Ronald J. Gutmann and Jian-Qiang Lu

CONTENTS

2.1 INTRODUCTION

Silicon complementary metal oxide semiconductor (CMOS) integrated circuits (ICs) continue to achieve increasing levels of integration for digital applications such as microprocessors and high-performance application-specific integrated circuits (ASICs). However, manufacturing cost and copper low-k interconnect delay may constrain future scaling past the 22 nm technology node. In addition, integration of high-performance, highly integrated digital ICs with other technologies such as analog/mixed-signal ICs, imagers, sensors, and wireless transceivers can be limited by packaging technologies, both conventional planar (2D) technologies and more recent three-dimensional (3D) stacking (either packaged die, bare die with full wafer thickness, or thinned die). While 3D packaging solutions offer increased functional density, electrical performance and interstrata interconnectivity are limited by wire bonding at the edge of the die. Micron-sized through-silicon vias (TSVs) offer significantly increased interstrata interconnectivity, along with significantly decreased interconnect parasitics (particularly inductance). By incorporating monolithic processes of wafer-to-wafer alignment, wafer bonding, wafer thinning, and interwafer interconnection, novel high-performance designs and unique system architectures become possible, combined with low manufacturing costs inherent with full-wafer IC processing.

This chapter is organized into five major sections, each including a summary perspective followed by a specific design or system architectural example. First, alternative wafer-level three-dimensional (3D) technology platforms are presented, with emphasis on dielectric adhesive bonding using benzocyclobute (BCB). Second, digital applications such as microprocessors and ASICs are discussed, with emphasis on performance prediction of static random access memory (SRAM) stacks for high-density memory. Third, analog/mixed-signal applications are discussed, with emphasis on wireless transceivers for software radios. Fourth, unique system architectures enabled by wafer-level 3D integration are discussed, with emphasis on both (1) point-of-load (PoL) DC/DC converters for power delivery with wide control bandwidth and (2) a novel system architecture using optical clocking for synchronous logic without latency. Fifth, technology and system drivers that need to be addressed before wafer-level 3D technology platforms are implemented for high-volume manufacturing are projected.

The authors believe that many unique system architectures will be realized with wafer-level 3D integration and hope that this chapter motivates IC system architects to extend their horizons to include 3D integration with low-parasitic interstrata interconnects. Two edited books on 3D integration were published in 2008, with contributions from many research groups in academia and industry [1, 2]. They provide comprehensive coverage of all high-performance 3D integration technologies and applications.

2.2   WAFER-LEVEL 3D TECHNOLOGY PLATFORMS

The purpose of 3D integration is to vertically stack and interconnect devices and circuits to form multifunctional, high-performance systems. The various wafer-level 3D technology platforms that have been investigated can be classified as (1) front-end-of-the-line (FEOL) platforms, (2) back-end-of-the-line (BEOL) platforms, and (3) wafer-level packaging (WLP) platforms. In FEOL-based platforms, strata of Si device layers are formed early in the IC process, by techniques such as epitaxial overgrowth and recrystallization of polysilicon. In BEOL-based platforms, active device layers are formed by wafer-to-wafer alignment and wafer bonding of two separately fabricated functional wafers, followed by wafer thinning and interwafer interconnect processing. In WLP-based platforms, fully processed wafers are aligned and interconnected using packaging-based processes such as flip-chip soldering.

FEOL platforms offer the highest density of interstrata interconnects and minimize increased complexity of IC interconnect processing or packaging. The potential for unique cell structures for digital designs is attractive, interstrata vias of 100 nm are feasible at technology nodes below 22 nm, and no new interconnect or packaging processes are required. However, high-quality devices in upper layers of the strata have not been achieved to date, primarily because of FEOL thermal budgets. FEOL platforms have been demonstrated for SRAMs and used in commercial products for nonvolatile memories (NVMs). However, the potential for heterogeneous integration, especially with different semiconductors requiring heteroepitaxy growth, remains a tremendous challenge. The future of FEOL 3D platforms is limited by the lack of a materials technology for achieving multiple strata with high-quality single-crystal silicon.

BEOL platforms offer a very high density of interstrata interconnects and allow fully interconnected device wafers to be processed similar to conventional planar 2D ICs. Interstrata vias of 1 μm and below are achievable today, with a variety of platforms outlined later in this section. Both heterogeneous integration of wafers with diverse technologies and high-density digital 3D stacks are feasible, both present and near-term technologies such as imagers with pixel processing and emerging technologies such as described in the last section of this book. Most research and development into 3D integration during the past decade has focused on BEOL platforms.

WLP-based platforms offer a more modest density of interstrata interconnects, while offering similar advantages for heterogeneous integration as BEOL platforms. Interstrata vias of 20–40 μm are achievable using micro flip-chip solder bumps. The via diameter limits the performance for advanced digital designs that require a large density of interstrata interconnects. While useful with 65 nm technology node CMOS digital ICs, the technology may not be scalable at the 22 nm node and beyond.

BEOL platforms, the focus of the remainder of this section, can be classified in many ways. Wafer bonding is either oxide-to-oxide, metal-to-metal (mostly copper-to-copper or Cu-to-Cu), or with dielectric adhesives (such as polyimides or BCB). The wafers to be bonded can be either fully interconnected (with 4–10 interconnect layers) or include only a local interconnect (silicide, salicide, or tungsten) with extensive interconnect levels after wafer bonding. The bonding process can form the interwafer via directly, such as with Cu-to-Cu bonding (via-first process flow), or wafer bonding can occur with blanket films on each wafer surface, such as bonding oxide-to-oxide or with dielectric adhesives, followed by a damascene process for an interstrata interconnect (via-last process flow). A handling wafer can be used to enable a back-to-front bond, with wafer thinning occurring on the handling wafer, or the wafers can be directly bonded to simplify the process flow and form a face-to-face bond. These alternative process flows and the enabling unit processes are described elsewhere.

Platforms based upon oxide-to-oxide bonding have been proposed by K.W. Guarini and colleagues at IBM, using a glass handling wafer to simplify wafer-to-wafer alignment, a via-last process flow, and face-to-back bonding with wafers processed through local interconnects [3]. Thus, the requirement of surface roughness to within 1 nm and surface planarity required for BEOL-compatible oxide bonding can be obtained. A modified version of this approach has been recently adapted by C. Keast and colleagues at Lincoln Laboratory in establishing a research-based 3D foundry capability [4]. These approaches with bonding occurring after local interconnects are formed in each wafer are suitable for integrated device manufacturers (IDMs). However, P. Enquist and colleagues at Ziptronix are applying oxide-to-oxide bonding after each wafer has a multilevel interconnect structure, both for wafer-to-wafer and die-to-wafer platforms [5]. Issues such as oxide surface roughness, planarity after multilevel interconnect processing, and subsequent bond strength need further evaluation.

The fundamentals of copper-to-copper bonding have been explored by R. Reif and colleagues at MIT, using a handling wafer, a via-first process flow, and face-to-back bonding [6, 7]. This group has demonstrated copper grain growth across the bonding interface at BEOL-compatible temperatures. P.R. Morrow and colleagues at Intel have demonstrated the compatibility of Cu-to-Cu bonding with TSV density suitable for microprocessors using face-to-face bonding without a handling wafer [8]. R. Patti and colleagues at Tezzaron are in pilot scale manufacturing of both stacked memories and memories stacked with a compatible processor wafer [9]. With such Cu-to-Cu bonding, the field dielectric is recessed prior to bonding in order to ensure the intimate contact of the Cu surfaces (Cu via and bonding pad) to be bonded as well as the high down force required for Cu bonding to be applied only to the Cu vias. A seal ring is probably needed around each die to avoid moisture interacting with the copper TSVs.

Dielectric adhesive bonding using polyimide and a via-last process flow have been investigated by C. Keast and colleagues at Lincoln Laboratories [10], P. Ramm and colleagues at Fraunhofer Institute [11], and M. Koyanagi and colleagues at Tohoku University [12] using polyimide adhesives, a via-last process flow, and face-to-face bonding without handling wafers. R.J. Gutmann, J.-Q. Lu, and colleagues at Rensselaer [13, 14] have established a viable process flow using BCB, which is widely used as a passivation layer, interlevel dielectric (ILD), and microelectromechanical system (MEMS) bonding material. Besides the traditional application in a via-last process flow, a via-first process flow has recently been established using Cu damascene-patterned vias with partially cured BCB to simplify the process flow [15, 16]. Compared to Cu-to-Cu bonding, this via-first platform has higher bond strength (due to the BCB bonding) and will not require seal rings around each die in most applications (as BCB-to-BCB contact is maintained across the bonding interface), at the expense of increased wafer planarity requirements.

To illustrate the status of these technology platforms, the via-last and via-first platforms with BCB adhesive bonding are described in more detail. A three-wafer stack of the via-last 3D platform is depicted in Figure 2.1, with summary comments describing the four key unit processes listed in Table 2.1. Partially cured BCB has been used in the bonding process to improve wafer-to-wafer alignment reproducibility with high bonding strength. Bonding of silicon to glass wafers, approximately thermal coefficient of expansion (TCE) matched to silicon, allows optical inspection of bonding defects, both immediately after bonding and after silicon thinning by grinding, chemical-mechanical polishing (CMP), and if desired, selective wet etching to an oxide layer using trimethyl ammonium hydroxide (TMAH). Collaborations with the University at Albany have demonstrated the capability for low-resistance interstrata interconnects with such a process flow [14].

In collaborations with both SEMATECH (wafers having two-level copper interconnect test structures with either oxide or low-k ILDs) and Freescale (wafers having four-level copper interconnects and fully functional 130 nm technology node silicon-on-insulator (SOI) CMOS devices), the robustness on the BCB bonding and three-step thinning process on advanced interconnect structures and CMOS devices and circuits has been demonstrated [17, 18]. A cross section of a CMOS test die with four

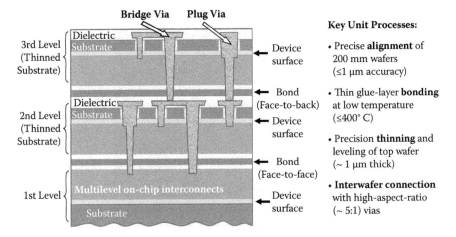

FIGURE 2.1 Schematic cross section of three-stratum stack of via-last 3D platform with BCB wafer bonding, showing bonding interface and vertical interstrata vias.

TABLE 2.1

Summary of Four Key Unit Processes

Key Unit Processes

- Precise **alignment** of 200 mm wafers (≤1 mm accuracy)
- Thin glue-layer **bonding** at low temperature (≤400°C)
- Precision **thinning** and leveling of top wafer (~1 μm thick)
- **Interwafer connection** with high-aspect-ratio (~5:1) vias

---

levels of Cu interconnect that has undergone two bonding and thinning processes and ashing of the initial BCB is shown in Figure 2.2; only the four-level interconnect, the ~150 nm SOI layer, and part of the buried oxide (BOX) layer remain, bonded to a prime Si wafer. Moreover, bonding integrity is maintained after liquid-to-liquid thermal shock (LLTS), high-temperature high-humidity (120°C and 100% relative humidity) conditions, and standard dicing and packaging operations [18, 19].

Recently we established a via-first process that provides direct electrical interconnects during the bonding operation, while achieving BCB-BCB bonding in the field region. A three-wafer stack of the via-first 3D platform is depicted in Figure 2.3. Vias and redistribution layers are formed by copper damascene patterning with partially cured BCB, with a cross section of the bonded interface shown in Figure 2.4. Excellent BCB-to-BCB interfaces are apparent, and good Cu-specific contact resistance has been demonstrated [15, 16]. Planarity requirements are stringent, but appear to be reasonable [20]. Further research and development is necessary to fully evaluate this promising platform.

In addition to these platforms using Cu or W interstrata interconnects with various process flows, several groups have explored capacitive interstrata coupling [20] with

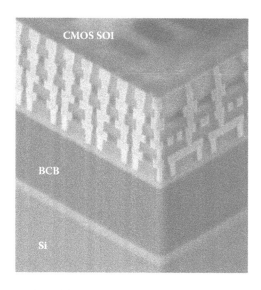

FIGURE 2.2   Cross section of CMOS SOI test structure with four-level copper interconnect after two bonding and thinning processes and BCB ashing.

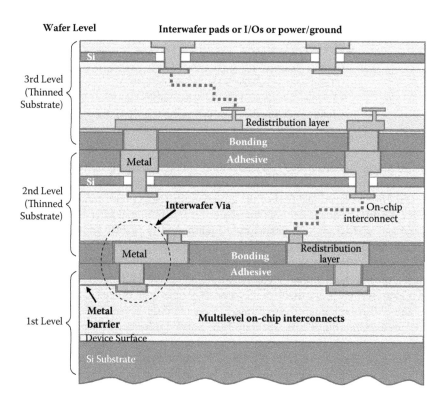

FIGURE 2.3   Schematic cross section of three-stratum stack of via-first 3D platform with patterned metal/adhesive (e.g., Cu/BCB) bonding.

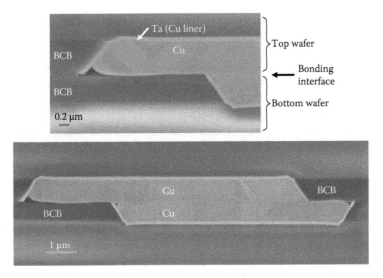

FIGURE 2.4 Cross-sectional focused ion beam (FIB)/SEM image of bonded damascene-patterned Cu/BCB wafers for via-first 3D platform, showing well-bonded Cu-to-Cu, BCB-to-BCB interfaces.

face-to-face bonding or inductive interstrata coupling [21] with either face-to-face or face-to-back bonding for noncontacting interstrata signal interconnects. While processing complexity is reduced, these alternatives may be limited to radio frequency (RF)/microwave systems with limited signal bandwidth. The application to high-speed digital ICs is more questionable, since the interconnect impedance is frequency dependent. Moreover, power distribution to and grounding between strata require interwafer electrical connections. These noncontacting techniques appear limited with BEOL-based platforms, but could be attractive with WLP-based 3D platforms.

While appreciable development is needed to fully evaluate these BEOL-based technology platform alternatives, no technology showstoppers have been demonstrated to date. The technology platforms that become selected for high-volume IC manufacturing will depend upon the following: (1) the two key application drivers (specifically, high-density digital ICs and heterogeneous integration of diverse process flows), (2) process compatibility (such as thermal and mechanical stress) with FEOL processing, die stack packaging, and diverse application environments, and (3) manufacturing metrics such as die yield, reliability, and cost.

2.3 DIGITAL IC APPLICATIONS

Digital ICs have been the key driver for CMOS scaling, and digital CMOS performance optimization drives the International Technology Roadmap for Semiconductors (ITRS). Analog/mixed signal, RF/microwave, sensors (including imagers), and power ICs leverage the process development and digital CMOS devices as much as possible. As mentioned previously, reduction of interconnect delay may be a key driver for 3D IC integration past the 22 nm technology node. Certainly, any

3D integration platform must be compatible with scaled digital CMOS memory, microprocessors, ASICs, and digital signal processors (DSPs) in order to achieve high-volume manufacturing.

The low-hanging fruit for 3D integration is memory stacks, such as dynamic random access memory (DRAM), SRAM, and flash. Both the regular structure of memories and compatibility with redundancy to achieve high within-wafer yield are desirable attributes for wafer-level 3D integration. In addition to a higher memory size proportional to the number of strata, the electrical performance (both access time and power dissipation) can be enhanced, as discussed later in this section.

More aggressive 3D integration includes processor and memory stack(s) for logic ICs such as microprocessors, ASICs, DSPs, and graphics chips. These planar ICs currently consist of large die often with large embedded memory, which can be partitioned into smaller die and realized with simpler process flows in multiple strata. In each case, die yield of the individual wafers will be increased, compensating yield loss in the additional processes for 3D integration described in the previous section.

An example of the performance advantages of 3D integration for SRAMs used for a 16 Mbyte L2 cache for real-time image processing has been analyzed at Rensselaer. An analytical model of the access time, cycle time, and dynamic power dissipation with circuit models confirmed using Cadence Spectre simulations. The analytic model, named PRACTICS (short for Predictor of Access Time and Cycle Time for Cache Stack), calculates performance parameters for all cache configurations with design parameters either fixed or variable within a certain range. Design parameters include number of strata, number of banks, associativity, word line parameters, and bit line parameters. Cycle time and dynamic power dissipation can be minimized by varying an evaluation parameter, lambda (equal to 0.0 to minimize cycle time, 1.0 to minimize dynamic power, and 0.5 for equal weighting) [22].

An optimized (with lambda = 0.5) performance comparison using 130 nm technology node CMOS technology (the most aggressive technology that could be fully evaluated with device and circuit parameters implemented in PRACTICS) is shown in Table 2.2 for three implementations: planar 2D, two-wafer stack, and four-wafer stack [22]. The four-wafer stack reduces the number of repeaters in the critical I/O channels by more than 80% (1344 to 224), decreases access time by 58% (5.09 to 2.04 ns), and achieves a cycle time of 0.5 ns, which provides real-time decoding of fast-acting videos without enhanced decoding algorithms. In addition, 3D implementations provide more uniform delay distribution with increasing number of strata, which improves pipelining efficiency.

While such memory stacks are very attractive in both embedded and stand-alone implementations, digital products such as microprocessors, ASICs, DSPs, and SoCs require integration of logic functions. While embedded SRAM and DRAM are becoming more of the die area for all CMOS ICs at newer technology nodes, high-performance processor cores are a particular concern in 3D integration. The interconnectivity is less regular, thermal concerns (already a major issue in 2D microprocessors) are enhanced, and prime power distribution can become more complicated. In a flip-chip packaged 3D die stack, heat sinking is through the bottom wafer in the original (i.e., before flipping) stack; the thermal path is lowest for this die. In a single-core processor, the wafer

TABLE 2.2
Performance Comparison of 2D and 3D Implementations for 130 nm Technology Node, 16 Mbyte L2 Cache (lambda = 0.5)

	Original Design	Original Configuration 3D	Optimized 3D
Layer	1	3	3
NumBank	8	8	16
Ndwl/Ndbl/Nspd	8/8/8	8/8/8	64/16/8
Address-in routing delay (ns)	1.21	0.09	0.13
Decoder delay (ns)	1.01	1.01	0.80
Wordline delay (ns)	1.00	1.00	0.40
Bitline delay (ns)	1.12	1.12	0.98
Output driver delay (ns)	0.56	0.56	0.52
Data output delay (ns)	0.90	0.10	0.12
Access time (ns)	5.80	3.88	2.96

with the processor core is in contact with heat sink, but for multiple-core processors alternative 3D implementations need to be considered (i.e., all cores in the wafer to be contacted to the heat sink or processor cores in all strata).

The electrical performance advantage for processors with 3D integration has also been analyzed as well. To first order for SoCs, the clock frequency increases by $N^{3/2}$ and the interconnect power decreases by $1/N^{1/2}$ [23, 24]; therefore, in a four-stratum implementation, the clock frequency increases by a factor of 8 and the interconnect power is reduced by a factor of 2. This first-order model result indicates the significant enhancement that can be potentially achieved in SoCs, but is probably somewhat optimistic. Further discussion is beyond the scope of this chapter.

2.4 ANALOG/MIXED-SIGNAL APPLICATIONS

Analog/mixed-signal ICs do not have the same scaling advantages as digital ICs. In fact, analog design issues such as dynamic range and power handling have become more difficult to achieve using a CMOS digital IC process at technology nodes below 130 nm. System-on-a-chip (SoC) realizations become more complex as MOS field-effect transistor (MOSFET) sizing and bias requirements for analog and RF/microwave applications increasingly deviate from digital CMOS devices at decreasing technology nodes. Wafer-level 3D provides an opportunity to partition analog and digital functions into different strata, with process flows and parameters selected to optimize each functional requirement. For example, Si-based microwave transceivers above 30 GHz usually incorporate silicon-germanium (SiGe) heterojunction bipolar transistors (HBTs) because of their higher cutoff frequency compared to Si MOSFETs. Incorporating baseband DSPs in high-performance digital CMOS in a BiCMOS process for a single-chip solution is not feasible (at best, the approach is not economical). However, a SiGe transceiver can be bonded to, and interconnected with, a Si CMOS DSP, forming a 3D-stacked SoC. In less cost-sensitive applications demanding highest electrical performance, higher-cost GaAs-based or InP-based HBTs can also be used in the transceiver stratum.

A long-range objective is to achieve a software radio, which requires an analog-to-digital converter (ADC) well beyond the current state of the art. A conceptual three-stratum future implementation is depicted in Figure 2.5a. Three strata

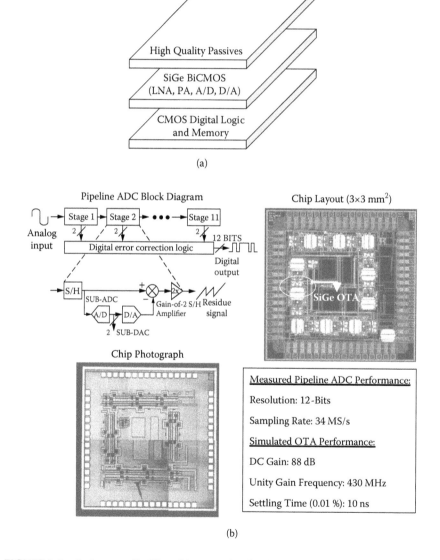

FIGURE 2.5 Software radio 3D architecture showing (a) three-die stack of SiGe BiCMOS transceiver and ADC with CMOS processor/memory and passive components and (b) prototype BiCMOS pipelined ADC.

are indicated: a top-surface passive-component-only layer with an on-die antenna and high-Q passives, a SiGe BiCMOS layer including a transceiver and analog-to-digital converter (ADC), and a digital CMOS processor and memory layer. The top layer could contain high permeability, low-loss magnetic thin films for compact, high-Q inductors, and high-k dielectric films for compact, high-Q capacitors, as well as a compact on-die antenna [25]. A key need in a true software radio in the long term or an effective software radio covering multiple frequency bands in the short term is the ADC.

We designed and evaluated a conventional pipelined ADC using gain-of-2 sample/hold (S/H) amplifiers with an operational transconductance amplifier (OTA) in a negative feedback loop using precise-value capacitors. An IBM 6HP process providing 47 GHz SiGe HBTs and 250 nm CMOS was used to design a wide-bandwidth, high-gain, fast-settling OTA using SiGe NPN HBTs in place of the usual NFETs in a cascade configuration. The 34 MS/s sampling rate with 12-bit resolution was limited by capacitive mismatch and lack of self-calibration techniques. An improved OTA with a triple cascade architecture should achieve a 115 MS/s sampling rate with 12-bit resolution; with digital self-calibration using a 7-bit pipeline seed, 205 MS/s is predicted, corresponding to an ADC figure of merit (FoM) of 4000 GHz/W for the 6HP process introduced in 2000 [25]. A microphotograph of the prototype ADC is shown in Figure 2.5b. In comparison, the ITRS predicts that a CMOS ADC will achieve such a FoM in 2012. While SiGe BiCMOS ADCs are more expensive than CMOS ADCs in a stand-alone product, such designs are attractive when integrated with a SiGe BiCMOS transceiver.

Other analog/mixed-signal architectures will be more readily realized in 3D with TSV interstrata interconnects, whether the technologies are Si MOS/Si CMOS, Si BiCMOS/Si CMOS, GaAs/Si CMOS, InP/Si CMOS, or any innovative 2D planar IC (such as a nanotechnology memory) with Si CMOS. Wafer-level 3D extends the reach of digital CMOS into monolithic stacked ICs with embedded DSP and processor capabilities. One principal constraint on SoC implementation is eliminated with such 3D platforms.

2.5 UNIQUE SYSTEM ARCHITECTURES

While SoC stacks of conventional architectures discussed in Section 2.4 are currently being pursued as demonstration vehicles with near-term product potential, future drivers of wafer-level 3D technology with TSVs for interstrata interconnect may be novel system architectures not envisioned with planar 2D ICs. We have explored two such innovations: (1) unique prime power delivery systems using a cell-based point-of-load (PoL) DC/DC converter [26] and (2) optical clock delivery across dies to maintain synchronous logic with low latency [27]. Both examples are briefly described in here.

Prime power delivery to ICs such as microprocessors, ASICs, and DSPs is an increasing system-level concern. A 3D stack with low parasitic interwafer interconnects enables DC/DC converters to be distributed across one stratum. Power is delivered to the packaged 3D die stack at relatively high voltage and to the signal electronics in a cell-based manner as depicted in Figure 2.6. From a power systems perspective, the point-of-load (PoL) converter is within a few millimeters of the load

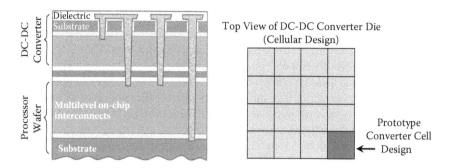

FIGURE 2.6 Cell-based 3D power delivery with DC/DC converter wafer.

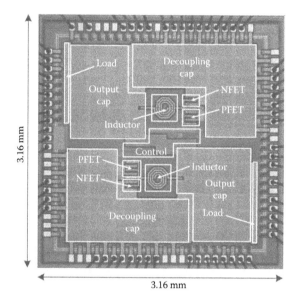

FIGURE 2.7 Microphotograph of PoL DC/DC buck converter fabricated through MOSIS.

with low parasitic interconnection. A large control bandwidth (~200 MHz) can be obtained, enabling dynamic power delivery to signal electronics with low transient time. With power input to a die-stack package at relatively high voltage, the large number of power and ground pins required for microprocessor and SoC packages can be significantly reduced.

A fully monolithic interleaved buck converter with linear feedback control was designed in a 180 nm SiGe BiCMOS process to operate at 200 MHz switching frequency and deliver 500 mA output current at 1.0 V. A microphotograph of the test chip that contains a MOSFET dummy load to simplify testing is shown in Figure 2.7 [26]. Even though the PMOS control switch has a total gate width of 16.6 mm and the NMOS synchronous switch has a total gate width of 11.0 mm, the die area is mostly used by input decoupling capacitors (31%), output capacitors (27%), and bond pads and electrostatic discharge (ESD) protection (31%). The low conversion efficiency of 64% obtained is limited by the available monolithic inductor and MOSFET dissipation in this first-generation prototype.

Scaling calculations indicate that this basic converter design with interleaved cells requires a die area of 250 mm² to power the Intel Core Duo processor [26]. While additional design effort is needed to achieve the 80–85% desired, the wide control bandwidth possible (limited to 10 MHz with the linear feedback control in this initial prototype, but with ~100 MHz anticipated) will enable effective dynamic power delivery with advanced logic ICs such as multicore processors.

Optical interconnects is one technique to distribute clocks across a synchronous logic die in a manner to minimize latency. In conventional approaches using planar 2D ICs, optical waveguide fabrication must be integrated with digital CMOS fabrication. With a 3D implementation an optical clock can be distributed with an H-tree network to perhaps 16 sections of a processor die, with only photodetectors required in the digital CMOS strata. With 45° mirrors in the H-tree network to orient the optical clock vertically where desired, the optical vias can either be an optical waveguide or an optical beam, as depicted in Figure 2.8. While the optical waveguide approach confines the optical beams, the fabrication requirements are difficult to achieve. Fortunately, the optical beam approach has acceptable performance with a variety of materials and dimensions appropriate for BEOL-compatible processing and 3D platform requirements [27].

We envision that implementation of such unique architectures with a high volume density of electronic devices may be limited in the future by signal integrity issues rather than the cost of such integration. With further scaling of digital CMOS almost assured past the 22 nm technology node, increasing capability of analog/mixed-signal technologies, opportunities for unique system architectures, and the ability to integrate optical and nanotechnology planar ICs, the integration capability of wafer-level 3D technology platforms will be almost mind-boggling. The limitation imposed by signal integrity constraints could become a fundamental constraint, and the use of optical interconnects could alleviate electronic signal integrity constraints.

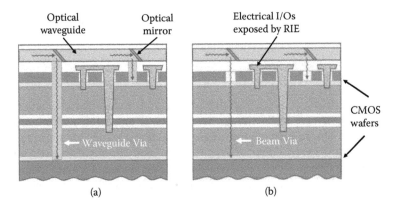

FIGURE 2.8 Clock distribution using 2D optical waveguide H-tree and 45° reflectors with (a) optical waveguide vias and (b) optical beam vias.

2.6 FUTURE DRIVERS FOR WAFER-LEVEL 3D IN IC MANUFACTURING

Present emphasis on FEOL device technology such as strained layers, high-k gate dielectrics, metallic gates, and wraparound gate structures is resulting in enhanced digital CMOS devices at the 45 and 32 nm technology nodes. By the 22 nm node these FEOL enhancements will result in a second-generation interconnect bottleneck (the first being in the 1990s when aluminum (Al) lines, tungsten (W) plugs as interlevel vias, oxide interlevel dielectrics (ILDs), and reactive ion etching (RIE) of Al as a patterning process were replaced with Cu lines and interlevel vias, low-k ILDs, and in-laid Cu as a patterning process (Damascene patterning). Wafer-level 3D is the only near-term alternative to planar (2D) ICs with Cu/lowest-k interconnects, as well as enabling heterogeneous integration of different planar technologies for innovative SoCs (at least further levels of system integration). The technology and infrastructure needs, which must be overcome before, or soon after, a decision to move to large-volume manufacturing, are discussed in this section, split into technology drivers, design drivers, equipment infrastructure drivers, and industry infrastructure drivers.

2.6.1 TECHNOLOGY DRIVERS

Selection of the product base to drive wafer-level 3D is paramount, as the technology requires establishment of additional infrastructure compared to 2D ICs. Digital CMOS with cell-to-cell interconnectivity offers the most design freedom, but requires the highest interstrata interconnectivity and probably the most involved design tools. Heterogeneous integration drivers such as mixed-signal SoCs and smart sensors (e.g., imagers with pixel processing) would impact selection of the technology platform. Certainly any 3D technology platform (BEOL or WLP based) offers different processing challenges and different opportunities. Perhaps a nanotechnology innovation to be integrated with digital CMOS will be such a driver, rather than a product that can be envisioned now. Down-selection of technology platform options will be needed prior to large-scale manufacturing.

2.6.2 DESIGN DRIVERS

A key design driver is the constraint imposed by wafer-level 3D integration, a common die size (as well as a common wafer size) for full utilization of the semiconductor area in each stratum. This constraint is somewhat limiting for an IDM or a 3D ASIC where die size is not constrained initially. However, for hard SoC implementation with some standard die being reused, a set of standardized die sizes is needed. In addition, design rules must be established that accommodate processing, electrical, and mechanical constraints. These will require simulation software as well to ensure signal integrity with semiconductor devices in multiple strata.

2.6.3 EQUIPMENT INFRASTRUCTURE DRIVERS

Wafer-level 3D integration requires new tools for IC processes, particularly wafer-to-wafer alignment and wafer bonding. While these tools have been available for many years from both EV Group and Suss Microtech and have been used in product manufacturing, they do not meet the requirements of the IC industry for high-volume 24/7 manufacturing. TSV technology is being developed for 3D packaging and can be readily modified for wafer-level 3D integration. In addition, the wafer bonding processes developed to date can take an hour or more; shorter process times are certainly desirable. Other process requirements are technology platform dependent, as described earlier.

2.6.4 INDUSTRY INFRASTRUCTURE DRIVERS

We are not certain whether wafer-level 3D technology platforms that align and bond wafers with complete interconnect structures will be integrated with the IC back-end or be incorporated into the expanding WLP infrastructure. Many major IDMs are actively involved in wafer-level 3D research and development; these IDMs would probably extend their BEOL technology base accordingly. We expect that major silicon foundries with multiple process flows will extend their BEOL technology base as well. However, second-tier silicon foundries with limited IC process flows will probably not extend their manufacturing base; major packaging foundries will probably extend their WLP and system-in-package (SiP) capabilities. Such industry infrastructure considerations could impact the acceptance of wafer-level 3D for high-volume mainstream products, at least in the near term.

2.6.5 PREDICTIONS

We believe that wafer-level 3D platforms (possibly die-to-wafer and probably wafer-to-wafer) will be driven by both high-speed digital CMOS at the 22 nm node and beyond and heterogeneous integration for SoCs (or to more closely achieve SoCs). We believe that this will occur within 3 to 5 years as (1) current FEOL technologies such as high-k gate dielectrics, metal gates, and wraparound gate structures become fully incorporated, (2) on-chip interconnect delay becomes a serious limitation to CMOS synchronous logic performance, (3) novel IC features and products become feasible in wafer-level 3D demonstrations, and (4) novel nanotechnology planar ICs are established that need to be integrated with digital CMOS processors.

ACKNOWLEDGMENTS

The authors gratefully acknowledge faculty colleagues, postdoctoral associates, and graduate students to the 3D interconnect research results: faculty contributors include Profs. T.S. Cale (technology platform), M. Hella (RF/microwave transceiver), J.F. McDonald (test mask design), P.D. Persans (optical interconnects), K. Rose (SiGe ADC and 3D memory performance prediction), and J. Sun (power delivery and DC/DC converter); visiting scientists and postdoctoral associates include A. Jindal (now at Micron), R.J. Kumar (now at Intel), K. Lee (now at Samsung), and F. Niklaus (on leave from KTH) (all in technology platform); graduate students include S. Devarajan (SiGe ADC) (now at Analog Devices), D. Giuliano (DC/DC converter) (now at MIT), Y. Kwon (technology platform) (now at Samsung), J.J. McMahon (technology platform) (at RPI), J. Yu (technology platform) (now at IBM), and A. Zeng (3D memory performance prediction) (now at Freescale). This research was funded principally through the Interconnect Focus Center, funded by MARCO, DARPA, and NYSTAR.

REFERENCES

1. P. Garrou, P. Ramm, and C. Bower, eds., *3-D IC Integration: Technology and Applications*, Wiley, New York, 2008.
2. C.N. Tan, R.J. Gutmann, and L.R. Reif, eds., *Wafer-Level Three-Dimensional (3D) IC Process Technology*, Springer, Berlin, 2008.
3. K.W. Guarini, A.W. Topol, M. Ieong, R. Yu, L. Shi, M.R. Newport, D.J. Frank, D.V. Singh, G.M. Cohen, S.V. Nitta, D.C. Boyd, P.A. O'Neil, S.L. Tempest, H.B. Pogge, S. Purushothaman, and W.E. Haensch, Electrical integrity of state-of-the-art 0.13 mm SOI CMOS devices and circuits transferred for three-dimensional (3D) integrated circuit (IC) fabrication, In *Digest of International Electron Device Meeting*, 2002, pp. 943–945.
4. J.A. Burns, B.F. Aull, C.K. Chen, C.-L. Chen, C.L. Keast, J.M. Knecht, V. Suntharalingam, K. Warner, P.W. Wyatt, and D.-R.W. Yost, A wafer-scale 3-D circuit integration technology, *IEEE Transactions on Electron Devices*, 53(10), 2507–2516, 2006.
5. P. Enquist, Room temperature direct wafer bonding for three dimensional integrated sensors, *Sensors and Materials*, 17(6), 307, 2005.
6. A. Fan, K.N. Chen, and R. Reif, Three-dimensional integration with copper wafer bonding, In *Proceedings of Electrochemical Society: ULSI Process Integration Symposium*, 2001, ECS PV 2001-02, pp. 124–128.
7. K.N. Chen, A. Fan, C.S. Tan, and R. Reif, Microstructure evolution and abnormal grain growth during copper wafer bonding, *Applied Physics Letters*, 81(20), 3774–3776, 2002.
8. P. Morrow, C.-M. Park, S. Ramanathan, M.J. Kobrinsky, and M. Harmes, Three-dimensional wafer stacking via Cu-Cu bonding integrated with 65-nm strained-Si/low-k CMOS technology, *IEEE Electron Device Letters*, 27(5), 335–337, 2006.
9. R. Patti, Three-dimensional integrated circuits and the future of system-on-chip designs, *Proceedings of the IEEE*, 94(6), 1214–1222, 2006.
10. J. Burns, L. McIlrath, C. Keast, A. Loomis, K. Warner, and P. Wyatt, Three-dimensional integrated circuits for low-power, high-bandwidth systems on a chip, In *Proceedings of IEEE International Solid-State Circuits Conference Technical Digest*, 2001, pp. 268–269.
11. P. Ramm, D. Bonfert, H. Gieser, J. Haufe, F. Iberl, A. Klumpp, A. Kux, and R. Wieland, InterChip via technology for vertical system integration, In *Proceedings of IEEE International Interconnect Technology Conference 2001 (IITC 2001)*, 2001, pp. 160–162.
12. K.W. Lee, T. Nakamura, T. One, Y. Yamada, T. Mizukusa, H. Hasimoto, K.T. Park, H. Kurino, and M. Koyanagi, Three dimensional shared memory fabricated using wafer stacking technology, In *Digest of International Electron Device Meeting*, 2000, pp. 165–168.
13. J.-Q. Lu, Y. Kwon, R.P. Kraft, R.J. Gutmann, J.F. McDonald, and T.S. Cale, Stacked chip-to-chip interconnections using wafer bonding technology with dielectric bonding glues, In *Proceedings of the 2001 IEEE International Interconnect Technology Conference (IITC)*, June 4–6, 2001, pp. 219–221.

14. J.-Q. Lu, K.W. Lee, Y. Kwon, G. Rajagopalan, J. McMahon, B. Altemus, M. Gupta, E. Eisenbraun, B. Xu, A. Jindal, R.P. Kraft, J.F. McDonald, J. Castracane, T.S. Cale, A. Kaloyeros, and R.J. Gutmann, Processing of inter-wafer vertical interconnects in 3D ICs, In *Advanced Metallization Conference in 2002 (AMC 2002)*, 2003, vol. V18, pp. 45–51.

15. J.J. McMahon, J.-Q. Lu, and R.J. Gutmann, Wafer bonding of damascene-patterned metal/adhesive redistribution layers for via-first 3D interconnect, In *Proceedings of the IEEE Electronic Components and Technology Conference*, 2005, pp. 331–336.

16. J.-Q. Lu, A. Jindal, Y. Kwon, J.J. McMahon, M. Rasco, R. Augur, T.S. Cale, and R.J. Gutmann, Evaluation procedures for wafer bonding and thinning of interconnect test structures for 3D ICs, In *2003 IEEE International Interconnect Technology Conference (IITC)*, June 2003, pp. 74–76.

17. R.J. Gutmann, J.-Q. Lu, S. Pozder, Y. Kwon, D. Menke, A. Jindal, M. Celik, M. Rasco, J.J. McMahon, K. Yu, and T.S. Cale, A wafer-level 3D IC technology platform, In *Proceedings of Advanced Metallization Conference*, 2003, pp. 19–26.

18. S. Pozder, J.-Q. Lu, Y. Kwon, S. Zollner, J. Yu, J.J. McMahon, T.S. Cale, K. Yu, and R.J. Gutmann, Back-end compatibility of bonding and thinning processes for a wafer-level 3D interconnect technology platform, In *2004 IEEE International Interconnect Technology Conference (IITC04)*, June 2004, pp. 102–104.

19. J.J. McMahon, R.J. Gutmann, and J.-Q. Lu, Three dimensional (3D) integration, In *Microelectronic Applications of Chemical Mechanical Planarization*, ed. Y. Li, John Wiley & Sons, New York, 2007.

20. Q. Gu, Z. Xu, J. Kim, J. Ko, and M.F. Chang, Three-dimensional circuit integration based on self-synchronized RF-interconnect using capacitive coupling, In *Digest of 2004 Symposium on VLSI Technology*, June 2004, pp. 96–97.

21. J. Xu, J. Wilson, S. Mick, L. Luo, and P. Franzon, 2.8 Gb/s inductively coupled interconnect for 3D ICs, In *Digest of 2005 Symposium on VLSI Circuits*, June 2005, pp. 352–355.

22. A.Y. Zeng, J.-Q. Lu, K. Rose, and R.J. Gutmann, First-order performance prediction of cache memory with wafer-level 3D integration, *IEEE Design and Test of Computers*, 22(6), 548–555, 2005.

23. J.D. Meindl, R. Venkatesan, J.A. Davis, J.W. Joyner, A. Naeemi, P. Zarkesh-Ha, M. Bakir, T. Mulé, P.A. Kohl, and K.P. Martin, Interconnecting device opportunities for gigascale integration (GSI), In *International Electronic Devices Meeting*, 2001, pp. 525–528.

24. J.W. Joyner and J.D. Meindl, Opportunities for reduced power dissipation using three-dimensional integration, In *2002 IEEE International Interconnect Technology Conference (IITC02)*, June 2002, pp. 148–150.

25. R.J. Gutmann, A.Y. Zeng, S. Devarajan, J.-Q. Lu, and K. Rose, Wafer-level three-dimensional monolithic integration for intelligent wireless terminals, *Journal of Semiconductor Technology and Science*, 4(3), 196–203, 2004.

26. J. Sun, J.-Q. Lu, D. Giuliano, P. Chow, and R.J. Gutmann, 3D power delivery for microprocessors and high-performance ASICs, In *Proceedings of 22nd Annual IEEE Applied Power Electronics Conference and Exposition (APEC 2007)*, 2007, pp. 127–133.

27. P.D. Persans, M. Ojha, R. Gutmann, J.-Q. Lu, A. Filin, and J. Plawsky, Optical interconnect components for wafer level heterogeneous hyper-integration, In *Materials, Technology, and Reliability for Advanced Interconnects and Low-k Dielectrics*, ed. R.J. Carter, C.S. Hau-Riege, G.M. Kloster, T.-M. Lu, and S.E. Schulz, 2004, vol. 812, pp. F6.11.1–F6.11.5.

3 Integration of Graphics Processing Cores with Microprocessors

Deepak C. Sekar and Chinnakrishnan Ballapuram

CONTENTS

3.1 INTRODUCTION

Power and thermal constraints have caused a paradigm shift in the semiconductor industry over the past few years. All market segments, including phones, tablets, desktops, and servers, have now reduced their emphasis on clock frequency and shifted to multicore architectures for boosting performance. Figure 3.1 clearly shows this trend of saturating frequency and increasing core count in modern processors. With Moore's law, on-die integration of many components such as peripheral control hubs, dynamic random access memory (DRAM) controllers, modems, and more importantly, graphics processors have become possible. Single-chip integration of graphics processing units (GPUs) with central processing units (CPUs) has emerged and also brought many challenges that arise from integrating disparate devices/architectures, starting from overall system architecture, software tools, programming and memory models, interconnect design, power and performance, transistor requirements, and process-related constraints. This chapter provides insight into the implementation, benefits and problems, current solutions, and future challenges of systems having CPUs and GPUs on the same chip.

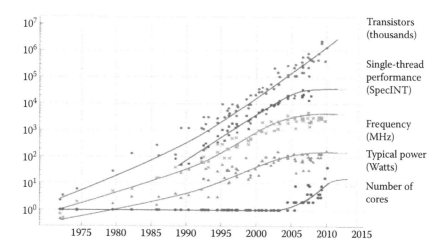

FIGURE 3.1   Microprocessor trends over the past 35 years.

3.2   WHY INTEGRATE CPUs AND GPUs ON THE SAME CHIP?

CPU and GPU microarchitectures have evolved over time, though the CPU progressed at a much faster pace, as graphics technology came into prominence a bit later than the CPU. Graphics is now getting more attention through games, content consumption from devices such as tablets, bigger-sized phones, smart TVs, and other mobile devices. Also, as the performance of the CPU has matured, additional transistors from process shrink are used to enhance 3D graphics and media performance, and integrate more disparate devices on the same die.

Figure 3.2 compares a system having a discrete graphics chip with another having a GPU integrated on the same die as the CPU. The benefits of having an integrated GPU are immediately apparent [1]:

- Bandwidth between the GPU and DRAM is increased by almost three times. This improves performance quite significantly for bandwidth-hungry graphics functions.
- Power and latency of interconnects between the CPU chip and GPU chip (of the multichip solution) are reduced.
- Data can be shared between the CPU and the GPU efficiently through better programming and memory models.
- Many workloads stress the GPU or the CPU and not both simultaneously. For GPU-intensive workloads, part of the CPU power budget can be transferred to the GPU, and vice versa. This allows better performance-power trade-offs for the system.

Besides these benefits, the trend of integrating GPUs with CPUs has an important scalability advantage. GPUs are inherently parallel and are known to benefit linearly with density improvements. Moore's law is excellent at providing density improvements, even though many argue that the performance and power improvements

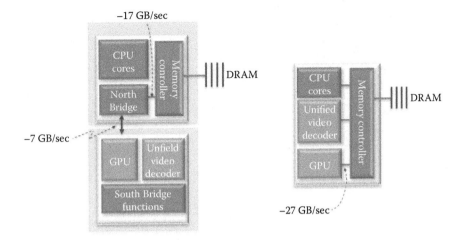

FIGURE 3.2   Left: A multichip CPU-GPU solution. Right: A single-chip CPU-GPU solution. (From Naffziger, in *Symposium on VLSI Technology*, 2011, pp. 6–10.)

it used to provide have run out of steam. By integrating GPUs, the scalability of computing systems is therefore expected to be better.

3.3   CASE STUDY OF INTEGRATED CPU-GPU CORES

In this section, we describe two modern processors, AMD Llano and Intel Ivy Bridge, which have both integrated CPUs and GPUs on the same die. These chips are often referred to as accelerated processing units (APUs).

3.3.1   AMD Llano [2]

The AMD Llano chip was constructed in a 32 nm high-k metal gate silicon-on-insulator technology. Figure 3.3 shows the integrated die that includes four CPU cores, a graphics core, a unified video decoder, as well as memory and I/O controllers.

The total die area is 227 mm². CPU cores were x86 based, with 1 Mbyte of L2 cache allocated per core. Each CPU core was 17.7 mm², including the L2 cache. Power gating was aggressively applied to both the core and L2 cache to minimize power consumption. A dynamic voltage and frequency scaling (DVFS) system was used that tuned supply voltage as a function of clock frequency to minimize power. Clock frequency was tuned for each core based on power consumption and activity of other CPU cores and the GPU. This was one of the key advantages of chip-level CPU and GPU integration—the power budget could be flexibly shared between these components based on workload and activity.

The GPU used a very long instruction word (VLIW) core as a basic building block, which included four stream cores, one special function stream core, one branch unit, and some general purpose registers. Each stream core could co-issue a 32-bit multiply and dependent ADD in a single clock. Sixteen of these VLIW

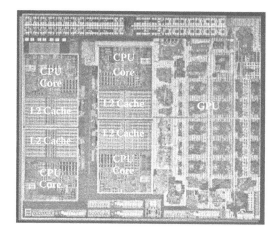
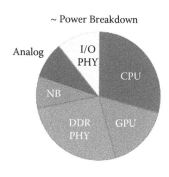

FIGURE 3.3 The 32 nm AMD Llano chip and a breakdown of its power consumption. I/O PHY and DDR PHY denote interface circuits for I/Os and DRAM, respectively, and NB denotes the Northbridge.

cores were combined to form a single-instruction, multiple-data (SIMD) processing unit. The GPU consisted of five such SIMDs, leading to a combined throughput of 480 billion floating point operations per second. Power gating was implemented in the GPU core as well, to save power. The GPU core occupied ~80 mm², which was nearly 35% of the die area. Power consumption of the GPU was comparable to that of the CPU for many workloads, as shown in Figure 3.3.

The CPU cores and the GPU shared a common memory in Llano systems, and a portion of this memory could be graphics frame buffer memory. As illustrated in Figure 3.4, graphics, multimedia, and display memory traffic were routed through the graphics memory controller, which arbitrated between the requestors and issued a stream of memory requests over the Radeon Memory Bus (RMB) to the North Bridge. Graphics memory controller accesses to frame buffer memory were non-coherent and did not snoop processor caches. Graphics or multimedia coherent accesses to memory were directed over the Fusion Control Link (FCL), which was also the path for processor access to I/O devices. The memory controller arbitrated between coherent and noncoherent accesses to memory.

3.3.2 Intel Ivy Bridge [3]

Ivy Bridge was a 22 nm product from Intel that integrated CPU and GPU cores on the same die. The four x86 CPU cores and graphics core were connected through a ring interconnect and shared the memory controller. Ivy Bridge had 1.4 billion transistors and a die size of about 160 mm². It was the first product that used a trigate transistor technology.

Figure 3.5 shows the system architecture of Ivy Bridge, where a graphics core occupied a significant portion of the total die. All coherent and noncoherent requests from both CPU and GPU were passed through the shared interconnect. The shared

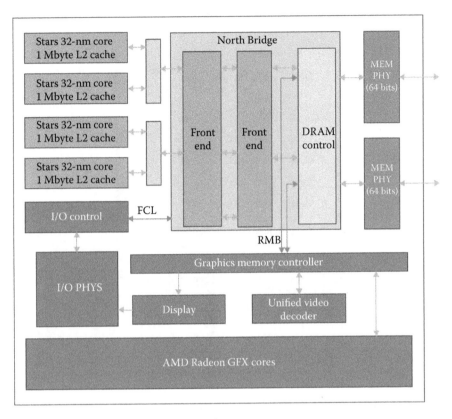

FIGURE 3.4 Block diagram of the AMD Llano chip. FCL, fusion control link; MEM, memory; PHY, physical layers; RMB, Radeon Memory Bus.

ring interconnect provided hundreds of GB/s bandwidth to the CPU and GPU cores. The last-level cache is logically one, but physically distributed to independently deliver data.

In Llano, coherent requests from the GPU went through a coherent queue and the noncoherent requests directly went to the memory. In Ivy Bridge, the CPU and GPU could share data in the bigger L3 cache, for example. The CPU could write commands to the GPU through the L3 cache, and in turn, the GPU could flush data back to the L3 cache for the CPU to access. Also, the bigger L3 cache reduced memory bandwidth requirements, and hence led to overall lower power consumption. Two different varieties of GPU cores were developed to serve different market segments. The graphics performance is mainly determined by the number of shader cores. The lower-end segment had eight shader cores in one slice, whereas the next-level segment had two slices. Different components of the processor were on different power planes to dynamically turn on/off the segments based on demands to save power. The CPU, GPU, and system agent were on different power planes to dynamically perform dynamic voltage frequency scaling (DVFS).

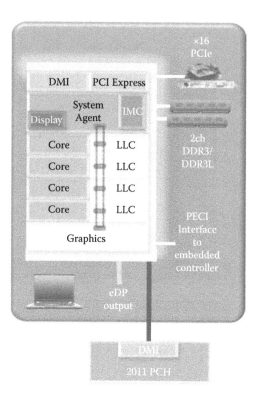

FIGURE 3.5   Block diagram of the Intel Ivy Bridge chip.

3.4   TECHNOLOGY CONSIDERATIONS [1]

The fundamentally different nature of CPU and GPU computations places interesting requirements on process and device technology. CPUs rely on using high-performance components, while GPUs require high-density low-power components. This leads to the use of performance-optimized standard cell libraries for CPU portions of a design and density-optimized standard cell libraries for GPU portions of a design. For example, the AMD Llano chip had 3.5 million flip-flops in its GPU, but only 0.66 million flip-flops in its CPU. The CPU flip-flops required higher performance and so were optimized differently. The flip-flop used for CPU cores occupied 50% more area than the flip-flop used for GPUs. The need for higher performance in CPU blocks led to the use of lower-threshold voltages and channel lengths in CPU standard cell libraries than in GPU ones.

The need for higher density in GPUs also leads to the requirement for smaller-sized wires than a pure CPU process technology. Smaller-sized wiring causes more wire RC delay issues since wire resistivity increases exponentially at smaller dimensions. This is because of scattering at sidewalls and grain boundaries of wires, as well as the fact that the diffusion barrier of copper occupies a bigger percentage of the wire area.

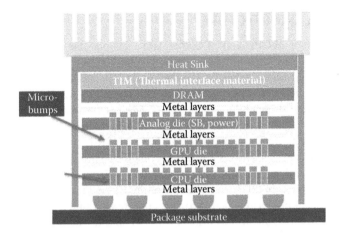

FIGURE 3.6 3D integration of CPU and GPU cores. (From Naffziger, in *Symposium on VLSI Technology*, 2011, pp. 6–10.)

In the long term, the differences in technology requirements for CPU and GPU cores could lead to 3D integration solutions. This would be particularly relevant for mobile applications where heat is less of a constraint. CPU cores could be stacked on a layer built with a high-performance process technology, while GPU cores could be stacked on a different layer built with a density-optimized process technology. DRAM could be stacked above these layers to provide the high memory bandwidth and low latency required for these systems. Figure 3.6 shows a schematic of such a system.

3.5 POWER MANAGEMENT

Most workloads emphasize either the serial CPU or the GPU and do not heavily utilize both simultaneously. By dynamically monitoring the power consumption in each the CPU and the GPU, and tracking the thermal characteristics of the die, watts that go unused by one compute element can be utilized by others. This transfer of power, however, is a complex function of locality on the die and the thermal characteristics of the cooling solution. The efficiency of sharing is a function of where the hot spot is and will vary across the spectrum of power levels. While the CPU is the hot spot on the die, for example, a 1 W reduction in CPU power could allow the GPU to consume an additional 1.6 W before the lateral heat conduction from CPU to GPU heats the CPU enough to be the hot spot again [1]. As the GPU consumes more power, it finally becomes the hot spot on the die, and the reverse situation occurs. A power management system that maintains repeatable performance must have sophisticated power tracking capability and thermal modeling to ensure maximum compute capability is extracted from a given thermal solution. Once that is in place, a chip with a CPU and a GPU can deliver far more computation within a thermal envelope than either design in isolation.

3.6   SYSTEM ARCHITECTURE

With the integration of CPU and GPU, there are a few possible system architectures, ranging from separate CPU/GPU memories to a unified one. Until recently, before CPU and GPU were on the same die, the CPU and GPU had their own memories and data were copied between these two memories for operation. AMD's roadmap shows progression from separate physical memory to hardware context switching for GPUs, wherein the chip will be able to decide which of its heterogeneous cores would best fit the needs of a particular application [4]. In 2011, AMD partitioned the physical memory into two regions, one each for CPU and GPU. The CPU paging mechanism was handled by hardware and the operating system (OS), while the GPU paging was handled by the driver. In 2013, AMD used a unified memory controller for CPU and GPU, and the 2014 platform added hardware context switching for the GPU. In contrast, Intel in 2011 used an on-die unified memory controller for CPU/GPU, including sharing the last-level cache between CPU and GPU. The on-die GPUs also have different levels of caches similar to CPU for texture, color, and other data.

The data transfer speed and bandwidth between CPU and GPU is critical for performance, and hence a scalable interconnect between these disparate cores is an important factor in APU design. Intel uses a ring interconnect to communicate between many CPU cores and the GPU. In addition to the interconnect design, other considerations in system architecture design include cache organization in both CPUs and GPUs, number of cache levels in the hierarchy, cache sizes at each level, cache policies, and sharing the last-level cache. AMD APU (Llano) uses two levels of CPU cache (L1 data cache is 64 KB, L2 cache is 1 MB 16-way) in each of four cores. There is no L3 cache, and the memory controller is shared between the CPU and GPU. In contrast, Intel's Sandy Bridge has an L1 data cache of 32 KB, L2 of 256 KB in each of four CPU cores, and inclusive L3 of 8 MB shared between four CPU cores and the GPU using a ring interconnect with a common memory controller behind the cache hierarchy. The decision to selectively determine the type of data to be cacheable/uncacheable and coherent/noncoherent between the CPU and the GPU can improve performance and bandwidth between cores in both types of system architecture design. Also, general purpose GPU (GPGPU) programming can take advantage of both the integrated cache and memory controller design to tap into computing power of GPUs. The bandwidth between key units in the system dictates the overall system performance.

3.7   PROGRAMMING AND MEMORY MODELS

Hardware manufacturers and software companies have been providing and supporting many tools and new languages to help parallel programming. These tools have evolved since the days when CPU and GPU were separate. First, let us look at a few tools provided by hardware manufacturers. We will then compare them with language extensions and standards proposed from Microsoft to write programs that support multicore CPUs, APUs, GPGPUs, and heterogeneous devices.

> **CUDA:** In early 2000, a few computer research labs built GPGPU application programming interfaces (APIs) on top of graphics APIs to enable and support GPGPU programming, and two such languages were

BrookGPU and Lib sh. Compute Unified Device Architecture (CUDA) [5] is Nvidia's approach to the GPGPU programming problem that lets programmers easily offload data processing to GPUs. The language is C with Nvidia extensions, like functions for GPU memory allocation, copy from host to device memory and back, as well as global and shared scope declarations. First, the data have to be copied to the GPU memory before GPU computation is invoked by the CPU, and the results are copied back to the main memory after GPU computation. The speedup is based on how efficiently the programmers code the parallelism. The CUDA architecture has evolved, and it currently supports many high-level languages and device-level APIs, such as OpenCL and DirectX. The integration of CPU and GPU on the same die will help ease memory bandwidth constraints.

OpenCL: Open Computing Language (OpenCL) [6] is a standard that provides a framework to parallelize programs for heterogeneous systems. Programs written using OpenCL can not only take advantage of multiple CPU cores and GPU cores, but also use other heterogeneous processors in the system. OpenCL's main goal is to use all resources in the system and offer superior portability. It uses a data and task parallel computational model and abstracts the underlying hardware. Data management is similar to CUDA, where the application has to explicitly manage the data transfer between the main memory and device memory.

HMPP: CUDA and OpenCL require programmers to rewrite their code in new language. Hardware Multicore Parallel Programming (HMPP) [7], on the other hand, provides a set of compiler directives that support multicore parallel programming in C. It is a flexible and portable interface for developing parallel applications that can use GPU and other hardware accelerators in the system. The HMPP directives divide the program into multiple codelets that can be run on multiple hardware accelerators. The HMPP runtime handles the parallel execution of the codelets, which have been translated earlier into a vendor programming model either by hand or with an available code generator. When the same code is run on a different system, the runtime HMPP will try to find the specified hardware accelerator and run it. If the specified accelerator is not available, the codelet will run on the host core. The HMPP directives also handle transfer of data between host memory and hardware accelerator memory. The HMPP model can also take advantage of shared memory systems like APUs and use fully synchronous execution between CPU and GPU. HMPP also supports Message Passing Interfaces and can run the codelet on a remote host with automatic generation of memory transfers.

AMP: Microsoft's C++ Accelerated Massive Parallelism (AMP) [8, 9] extension allows programmers to express data parallelism as part of the C++ language and lets compilers and DirectX tools create one binary and run on any heterogeneous hardware that supports it to improve performance. The C++ AMP understands if the data have to be copied to the GPU based on the memory model and is transparent to the programmer.

The complexity of programming CPU, GPU, and other accelerators is that memory models are different, ranging from weak, nonuniform memory access (NUMA), etc., from a hardware perspective to a software-managed memory model and recently proposed unified memory model for CPU/GPU and heterogeneous processors. The C++ AMP programming model is progressive looking and works well with AMD's Heterogeneous System Architecture (HSA) roadmap that plans to support programming and memory models to support efficient programming of APUs.

GCN: AMD's Graphics Core Next (GCN) is a new design to make GPUs capable of doing compute tasks equally well. AMD is moving from a VLIW graphics machine to a non-VLIW or SIMD machine as a basic block for GPU. In the new SIMD compute unit-based GPU, AMD plans to support high-level language features such as pointers, virtual functions, and support for GPU to directly call system services and I/O. Also, the new graphics architecture can serve all segments. In the client segment the new architecture provides similar power/performance as the VLIW machine. And in the server segments, the SIMD basic block with the right software tools can be used as an efficient compute engine and also take advantage of integrated CPU and GPU. The APUs will also provide a view of unified memory between both CPU and GPU to make the data communication efficient by eliminating copying of data between host and accelerator memory that is required in CUDA or OpenCL.

HSA: As the demand to use GPU for purposes other than graphics has been increasing, the GPU architecture is also evolving to support both graphics and compute engines. Also, there is an increased need of software support to use GPUs as compute engines in parallel with other asymmetric cores in the system. Heterogeneous System Architecture (HSA) provides an ecosystem to build a powerful system from combining simple, efficient, unique, and disparate processors. AMD supported coherent and unified memory for CPU and GPU in 2013, and provided GPU context switching in 2014. The HSA roadmap is to move from physical integration, architectural integration, and finally to system integration. The architectural integration supports unified address space, pageable system memory for GPU, and fully coherent memory between CPU and GPU. The system integration provides preemption, context switching, and quality of service (QoS). AMD's plan is to treat GPU as a first-class core and give equal privileges as CPU cores with Heterogeneous System Architecture. Currently, AMD is positioned well with its commitment and support to architecture, hardware, OS, tools, and applications for the HSA framework and ecosystem.

In summary, there are three main vectors in heterogeneous computing—Instruction Set Architecture (ISA), memory model, and programming model—that cover different memory models and ISAs. We can see that CUDA, OpenCL, HMPP, and C++ AMP target the programming model vector, Graphics Compute Next targets the ISA vector, while APUs themselves define the underlying memory model. Another important secondary vector and effect of memory models and programming models is data synchronization. The data synchronization between asymmetric cores depends on the memory model supported by hardware and the one used by the application. As more parallelism is exploited by different accelerators and hardware that support many core counts, the cost of synchronization increases probably to a point where the cost may exceed the benefit of the result from parallelism. We know that graphics cores use high memory bandwidth. In a system with APUs, if this high bandwidth traffic from GPU is coherent with the CPU cores, then the snoop bandwidth to all the CPU cores will be high, which will not only increase power and reduce performance of CPU cores, but also increase the latency of the GPU core. So, an efficient synchronization mechanism becomes important as we move toward programming APUs and future heterogeneous systems. As the potential for parallelism is increased, memory consistency restrictions on hardware may limit the performance.

3.8   AREA AND POWER IMPLICATIONS IN APUs

The area dedicated for graphics is increasing as mobile and desktop devices are demanding more media and graphics performance. In AMD's Llano APU, GPU occupies around 35% of the die area, while Intel's Sandy Bridge GPU occupies 20% of the die area. Also, the frequency at which the GPU operates is increasing and is taking more percentage of the total power budget, and hence efficient power management is required to increase the battery life. There may come an inflection point, where bigger control units like thread scheduler, dispatcher, scoreboard logic, register file management, and other control units will become a bottleneck and adding more parallel execution units in the GPU may not provide the expected benefit. Similar to the multicore CPUs, instantiation of multiple GPU units is a possible solution. But, the management of two GPU cores has to be supported by drivers, and this opens up new challenges for driver management.

3.9   GPUs AS FIRST-CLASS PROCESSORS

GPUs have matured over time with powerful graphics and compute engine capabilities. Also, programmers have many options to program APUs, and the tools range from language extensions and declarations/annotations in the programming language to aid compilers and low-level APIs with good debugging tools. Recent trends are to use on-board GPU clusters, networked GPU clusters, and GPU virtualization along with CPU clusters for many applications and creation of clouds and personal cloud networks. All these techniques still use CPU and device driver handshakes to communicate data between host and device memory. Though the GPUs are used for diverse purposes and applications, complete integration of CPUs and GPUs is missing due to lack of framework, standards, and challenges in making this happen. To extend GPU support to the next level, the OS needs to treat GPUs as first-class hardware for complete integration of CPUs and GPUs in the APUs.

To understand the requirements and challenges in making the GPU a first-class processor from the OS perspective, let us briefly look at the current method of interaction between the OS, applications, and CPU and GPU hardware. The main application runs on the CPU and fills up system memory with required information for graphics processing. The graphics part of the main application uses OpenCL, CUDA, or any other above-described programming model API to communicate with the device driver. The device driver in turn will send GPU commands to a ring buffer that acts as a queue to the underlying GPU hardware. Current GPU schedulers mostly use FIFO process scheduling that assumes a run-to-completion model and does not preempt tasks for intelligent scheduling. Some of the basic requirements for an OS, like preemption and task migration, are not supported in the current generation of GPUs.

Figure 3.7 captures the above-described flow between the main application running on a CPU that uses system memory and device drivers to run the graphics application on graphics hardware. Recent architectures allow the CPU part and the graphics part of the main application to share information directly in system memory that is transparent to the OS. The drivers, CPU, and GPU part of the application take care of data transfer between cores and coherency.

The number of shader cores and other necessary blocks is increased to handle more data and make the GPU more powerful. There comes a point where just increasing the number of cores may not be easily achieved; instead, adding another GPU core would be easier and complexity effective. When the number of GPUs has to be increased, the system architecture has to be redesigned. One solution is to add intelligence at the device driver level. Another solution is to add preemption for GPUs and provide control to the OS and treat GPUs as first-class processors [10–12]. With the addition of GPUs, the OS needs to take GPU execution times, resources, migration of GPU tasks, and other information into consideration to come up with new scheduling strategies accordingly. Also, the OS needs to consider different underlying architectures and binaries that will affect the way programs are loaded and dynamically linked. Another challenge is the migration of an already started task from the CPU core to the GPU core for acceleration will require saving the states from CPU cores and migrating them to GPU cores. Current GPUs provide support for indirect branching to make calls and save state information during exceptions. These primitives and support for shared virtual memory and unified address space proposed in future GPUs are steps in the right direction to make GPUs first-class processors from both a programmer and an OS perspective.

Figure 3.8 shows a future ecosystem, where CPUs and GPUs can communicate through a shared cache and shared memory system. The OS schedules the workloads to both CPUs and GPUs and treats them as equals.

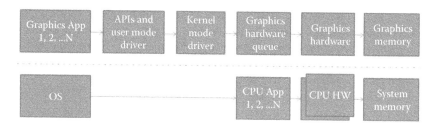

FIGURE 3.7   Interaction between OS, CPU, and GPU to run graphics application.

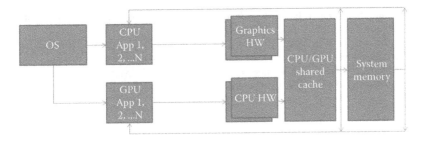

FIGURE 3.8 The OS treats both CPUs and GPUs as first-class processors.

3.10 SUMMARY

Moore's law has made the integration of CPUs and GPUs on the same chip possible. This has several important implications for process technology, circuit design, architecture, programming, and memory models, as well as software tools. In terms of process technology, GPUs prefer slower but higher-density libraries than CPUs, and require smaller-sized wires as well. This leads to separately optimized process technologies for CPU and GPU portions of a chip, and could eventually lead to 3D integration solutions where CPU and GPU portions of a chip can be stacked atop each other. Power budgets can be efficiently shared between CPU and GPU portions of a design based on thermal and power delivery considerations. Different memory sharing models and interconnect networks are possible for chips that integrate CPUs and GPUs, and performance can be quite sensitive to these decisions. Currently, there are different programming models based on memory models, ISAs, and synchronization mechanisms with high overhead. These constructs will ease the programming model and also help define simpler memory models that help portability and programmability of APU programs. Unified address spaces, preemption, task migration, and other constructs will enable the OS to treat GPUs as CPUs in the long term.

REFERENCES

1. S. Naffziger, Technology impacts from the new wave of architectures for media-rich workloads, In *Symposium on VLSI Technology*, 2011, pp. 6–10.
2. A. Branover, D. Foley, M. Steinman, AMD Fusion APU: Llano, *IEEE Micro*, 32(2), 28–37, 2012.
3. S. Damaraju, V. George, S. Jahagirdar, T. Khondker, R. Milstrey, S. Sarkar, S. Siers, I. Stolero, A. Subbiah, A 22nm IA multi-CPU and GPU system-on-chip, In *Proceedings of International Solid State Circuits Conference*, 2012.
4. P. Rogers, AMD Fusion Developer Summit, 2011.
5. http://www.nvidia.com/object/cuda_home_new.html.
6. http://www.khronos.org/developers.
7. http://www.openhmpp.org/en/OpenHMPPConsortium.aspx.
8. http://msdn.microsoft.com/en-us/library/hh265137(v = vs.110).aspx.
9. http://realworldtech.com/page.cfm?ArticleID = RWT062711124854.
10. S. Kato, K. Lakshmanan, Y. Ishikawa, R. Rajkumar, Resource sharing in GPU-accelerated windowing systems, In *Proceedings of the IEEE Real-Time and Embedded Technology and Applications Symposium*, 2011, pp. 191–200.

11. S. Kato, S. Brandt, Y. Ishikawa, R.R. Rajkumar, Operating systems challenges for GPU resource management, In *OSPERT 2011*, 2011.
12. T. Beisel, T. Wiersema, C. Plessl, A. Brinkmann, Cooperative multitasking for heterogeneous accelerators in the Linux completely fair scheduler, In *Proceedings of IEEE International Conference on Application-Specific Systems, Architectures, and Processors (ASAP)*, IEEE Computer Society, September 2011.

4 Electrothermal Simulation of Three-Dimensional Integrated Circuits

Shivam Priyadarshi, Jianchen Hu, Michael B. Steer,
Paul D. Franzon, and W. Rhett Davis

CONTENTS

4.1 INTRODUCTION

Three-dimensional integrated circuit (3D IC) is a promising technology that has potential to achieve higher device densities than technology scaling alone while improving energy efficiency. 3D ICs utilize vertical dimension for stacking different ICs. The vertical stacking largely reduces the total wire length and routing congestion compared to a conventional 2D implementation, and thus reduces interconnect delay and power consumption [1, 2]. Furthermore, 3D IC technology can broaden the horizon of what a system-on-chip can achieve by providing the capability to integrate disparate integrated technologies (such as technologies supporting radio frequency (RF) and high-performance logic devices) on a single chip. This type of heterogenous technology integration can significantly reduce the delay and power consumption [3, 4], and thus facilitates building energy-efficient systems. Moreover, even within one technology, different generations (for example, 45 and 32 nm logic CMOS) can be stacked to realize the cost-benefit from the better yield of the mature node [5].

To illustrate the 3D IC specific features, the first three tiers of a five-tier 3D technology, called FreePDK3D45 [6], are shown in Figure 4.1. Circuits in different tiers of a 3D IC can communicate with each other through different types of through-silicon vias (TSVs) and metal microbumps. For example, as shown in Figure 4.1, FreePDK3D45 supports three kinds of TSVs: (1) down via (VDN), (2) up via (VUP), and (3) through-tier via (VTT). Furthermore, this technology supports two types of microbumps: (1) front-side microbump, which is contact between the top metals of two tiers,

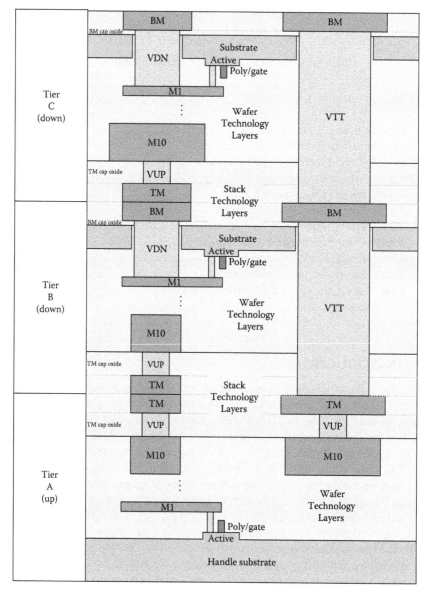

FIGURE 4.1   Cross section of the first three tiers of the FreePDK3D45 technology.

i.e., face-to-face bonding (e.g., bonding between tier A and tier B in Figure 4.1), and (2) back-side microbump, which is contact between the back metal of one tier and the top metal of another tier, i.e., back-to-face bonding (e.g., bonding between tier B and tier C in Figure 4.1). In this figure, tier A can be a digital logic technology, tier B can be a dynamic random access memory (DRAM) technology, and tier C can be an RF technology all integrated in a monolithic 3D die using TSVs and microbumps. This shows that the off-chip interconnects can be eliminated, which has potential to reduce overall delay and power consumption. Moreover, one of the major advantages of 3D ICs in realizing processor-based digital systems is capability to achieve high data bandwidth. For example, if tier A is a processor and tier B is a DRAM, then higher data bandwidth can be simply achieved by increasing the density of TSVs and microbumps. This is not possible in 2D implementation having off-chip interconnect between processor and DRAM because bandwidth is often limited by the possible number of pins, which is dictated by cost.

There are four key design challenges associated with 3D ICs that must be addressed for the widespread adaptation of this technology. These challenges are associated with (1) heat dissipation, (2) power delivery network design, (3) floor planning, and (4) design for test. This chapter is focused on thermal challenges of 3D ICs. Increased volumetric density in 3D ICs leads to higher power density. Furthermore, lower thermal conductivities of the intertier and intermetal dielectrics increase the thermal resistances and block the heat flow. These result in increased on-chip temperature, which can adversely affect performance (by means of mobility degradation), power (by means of exponential increase in leakage current), reliability (by means of electromigration, time-dependent dielectric breakdown, negative bias temperature instability, etc.), and cost (by means of increased cooling cost) of 3D ICs. Furthermore, the thermal issues in 3D ICs affect all the aforementioned design concerns, which makes the modeling of dynamic interaction between electrical and thermal characteristics more of an absolute necessity than ever before. For example, increased temperature in 3D stack can magnify the strength of the positive feedback loop between self-heating (i.e., joule heating) in the power grid and temperature-dependent electrical resistivity, which can significantly increase the IR drop in the power grid. Furthermore, the positive feedback loop between leakage power and temperature strengthened by increased temperature in the 3D stack can significantly change the power density and constrain the floor planning optimizations. The thermal induced TSV stress can affect the mobility of transistors in the proximity of TSVs, causing timing variations in 3D ICs, which makes the delay-fault testing more challenging. Thus, CAD tools and flows for modeling the dynamic interaction between electrical and thermal phenomenons, i.e., electrothermal simulation, are essential for successfully designing 3D ICs.

Now a question arises: At which level of design abstraction should electrothermal simulation be performed? The answer is that it should not be limited to only one design abstraction, but should be performed at different levels of design abstraction because this provides various opportunities for optimization at different design costs. For example, a study by LSI logic shows that power can be reduced by 20, 10, and 5% by optimizations at the register transfer level (RTL), gate level, and transistor level, respectively, whereas 80% reduction can be achieved at the electronic system level (ESL) [7]. At the system level, optimizations can be done with less effort than at the RTL, gate-level, and transistor-level abstractions. In system-level explorations attention is more on capturing the trend rather than absolute values. Optimizations at the RTL, gate level, or transistor level require more involved and detailed gate or circuit simulation for identifying the opportunities. However, simulation at these levels is still required before design sign-off for accurately estimating the temporal and spatial variations in temperature across the 3D stack. It is also required for determining the precise locations of hotspots and capturing the localized variations in the device parameters around the hotspots. This chapter describes an approach for system-level electrothermal simulation of 3D ICs.

This chapter is organized as follows. Section 4.2 presents major approaches to electrothermal simulation. An approach to system-level electrothermal simulation is presented in Section 4.3. Using the approach described here, a case study comparing a 2D and three 3D implementations of a quad-core chip multiprocessor (CMP) is presented in Section 4.4. Section 4.5 concludes this chapter.

4.2 ELECTROTHERMAL SIMULATION

The time dependent three-dimensional heat diffusion equation is

$$\rho c \frac{\partial T}{\partial t} = Q(x,y,z,t) + k(T) \left[\frac{\partial^2 T}{\partial x^2} + \frac{\partial^2 T}{\partial y^2} + \frac{\partial^2 T}{\partial z^2} \right] \tag{4.1}$$

where T is temperature, t is time, ρ is material density, c is specific heat, $Q(x, y, z, t)$ is the rate of heat generation, and k is temperature-dependent thermal conductivity.

The temperature profile can be obtained by solving Equation 4.1. In Equation 4.1, the heat generation rate $Q(x, y, z, t)$ is equivalent to the power consumption in an electrical system. So electrothermal simulation requires close interaction between electrical and thermal simulations. The electrical simulation is required to obtain information on power dissipation, which is fed to the thermal simulation. The thermal simulation is required to obtain temperature information, which is fed to electrical simulation, and temperature-dependent electrical parameters, such as leakage current, mobility, threshold voltage, etc., are updated accordingly. In dynamic electrothermal simulation, there are two approaches to model the coupling between electrical and thermal simulations: (1) relaxation and (2) direct method.

In the relaxation method, electrical and thermal simulations are performed separately with temperature updates passed from the thermal simulator to the electrical simulator and power updates passed from the electrical simulator to the thermal simulator [8, 9]. These updates occur at intervals much longer than the time duration of electrical transients, e.g., hundreds of clock cycles or more for a digital circuit. Typically in the direct method [11–13], an electrical circuit model of a thermal system is created based on the thermal-electrical analogies shown in Table 4.1 [10]. The electrical and thermal circuit models are solved simultaneously as if they were one large electrical circuit model. This effectively converts an electrothermal simulation to pure electrical simulation.

TABLE 4.1

Thermal-Electrical Analogy

Thermal	Electrical
Temperature T (K)	Voltage, V (V)
Heat, Q (J)	Charge, Q (C)
Heat transfer rate, q (W)	Current, i (A)
Thermal resistance, R_T (K/W)	Electrical resistance, R (V/A)
Thermal capacitance, C_T (J/K)	Electrical capacitance, C (C/V)
Temperature rise, $\Delta T = {}_q R_T$	Voltage difference, $\Delta V = iR$

The heat transfer rate (q) and electrical current (i) are functions of temperature (T) and voltage (V). The heat transfer rate corresponds to power dissipation in an electrical system. This method requires solving the following set of equations using iteration [12]:

$$
\begin{pmatrix} Y_E & 0 \\ 0 & Y_{TH} \end{pmatrix} \begin{pmatrix} V \\ T \end{pmatrix} = \begin{pmatrix} i(V, T) \\ q(V, T) \end{pmatrix}
\tag{4.2}
$$

In Equation 4.2, Y_E corresponds to the electrical modified nodal admittance matrix and Y_{TH} corresponds to the thermal admittance matrix.

The relaxation method is easier to implement as existing electrical and thermal simulators can be directly used, but accuracy of this method cannot be assumed in strongly coupled thermal problems [11]. Furthermore, very fast changes cannot be considered in this method [12]. The direct method requires a more complex physically consistent implementation than does the relaxation approach, but is capable of handling very fast changes [13]. In general, for large-scale simulations, relaxation methods are computationally more efficient than direct methods, but direct methods are more accurate than relaxation methods. In relaxation methods, a trade-off between simulation speed and accuracy can be done by changing the length of the interval after which the electrical and thermal simulators exchange their updates. Thus, relaxation methods are more suitable for system-level simulations, which require good computational efficiency in order to quickly explore a large design space. Direct methods are more suitable for gate-level simulations where good accuracy is essentially required.

4.3 SYSTEM-LEVEL ELECTROTHERMAL SIMULATION

3D integration provides a rich variety of choices to system designers such as homogenous vs. heterogeneous partitioning of a design, number of stacking layers in 3D design, integration of heterogeneous technologies, and different types of 3D bonding methods, resulting in vast 3D design space. The full exploitation of the benefits of 3D integration requires a system-level exploration flow that can facilitate in finding an optimal 3D design by thermally comparing possible early design choices, and thus eliminating thermally bad design early on. This can significantly reduce the development cost and risk. The thermal simulation of an integrated circuit requires some granularity of physical and power details based on which it can be categorized in two groups: (1) fine-grained and (2) coarse-grained simulation. A fine-grained thermal simulation requires accurate physical details, such as transistor-level layout, dielectric and interconnect material details (e.g., thermal conductivity, specific heat capacity, density, etc.), and power associated to each or a group of transistors. Lack of some of these details (e.g., layout, power associated to transistors) early in the design flow and enormous computational cost of fine-grained simulation prohibit its use in system-level design space exploration. Thus, at the system level, a coarse-grained thermal simulation flow that allows abstracting the physical details and estimating the power consumption at a higher level of design abstraction is required. In this section such a system-level electrothermal flow is presented.

4.3.1 SYSTEM-LEVEL FLOW

Figure 4.2 [14] shows a system-level 3D design exploration flow that contains two parts: the front end and the backend. In the front end, the designer creates a high-level system description (using component IP blocks) to evaluate power and performance and provide input to the backend. The transaction-level modeling (TLM) methodology is used to capture the dynamic effects (e.g., transient variations in power profile) resulting from the complex interaction between the system components (i.e., IP blocks) while running an application. A high-level power model is integrated with the TLM framework to calculate the transient power, average power, and performance based on IP configurations. The backend of the flow starts with a rough floor plan obtained from a system-level description using an area model of components. Users must specify technology information of each layer/tier of the stack (the wafer technologies), information about TSV-based stacking of different tiers (a stack technology), and material properties. To speed up the thermal simulation, layers in the wafer technologies are collapsed into a reduced set of layers, called composite layers. The composite layers, rough floor plan, and power information are fed into a thermal simulator to generate the static and dynamic temperature profiles. The system description can be easily changed based on obtained power, performance, and thermal results and a new iteration can be performed. This flow facilitates fast design space exploration to find the optimal 3D design without doing a detailed implementation of all the design choices. The following sections describe the components of the system-level flow in detail.

4.3.2 SYSTEM DESCRIPTION

As a first step, designers are required to create a high-level system description for a set of design parameters using functional and timing models of component IP blocks. This step includes two phases: IP configuration and logical partitioning. In the IP configuration phase, the designer sets appropriate parameters for each IP block (core type, cache size, etc.), and in the logical partition phase, the designer specifies how the IP blocks are connected to each other and how the data are transferred

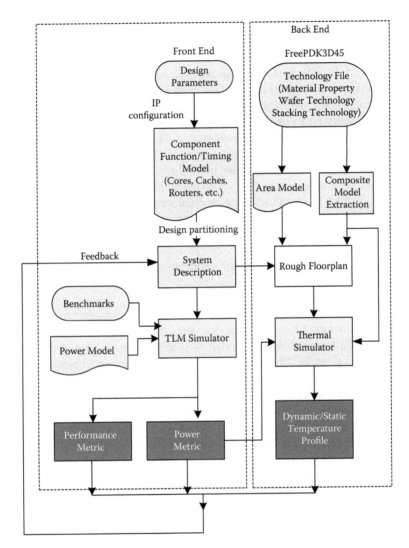

FIGURE 4.2 System-level CAD flow for 3D design space exploration. (From Priyadarshi et al., In *Proceedings of IEEE International 3D System Integration Conference*, February 2012, pp. 1–8.)

among the blocks. If the system has caches and memories, the cache policy and cache protocol are also determined in this phase.

4.3.3 Transaction-Level Modeling

It is important to capture the dynamic effects of an underlying application on power and performance, and to achieve this, the transaction-level modeling approach is used. Transaction-level modeling is a technique that separates the computation and communication of a system and hides the details not required at the early phases of the design flow, resulting in fast simulation. In a TLM representation, IP blocks contain concurrent processes that execute their behaviors, whereas communication is abstracted from cycle-by-cycle operation to an abstract operation called transaction. Communication is implemented as channels that hide protocols from the IP blocks, and a transaction is initiated by calling the interface functions of channels. TLM-based simulation can greatly reduce the simulation time, compared to register transfer level (RTL) simulation, with acceptable timing accuracy [16], and thus is suitable for system-level design space exploration. The authoring of TLM simulations is the most time-consuming part of this flow, and therefore users will most often need to reuse existing TLM frameworks. The TLM simulator used in the case study presented here is the GEMS multiprocessor simulator [17], which maintains a global event queue to manage all the blocks. Each block can schedule an event for the subsequent block. The global event queue triggers all the appropriate events scheduled for the current cycle. To ensure functional and timing correctness, all the outputs of a block are available to other blocks in the next cycle. Thus, all the blocks can be triggered out of order. The switching activity results from this event-driven simulator are extracted in terms of cycle counts.

4.3.4   POWER MODEL

A high-level power model is needed for estimating the power consumption. Computer architects typically use analytical models to predict the power. However, it can be difficult to tune these models to accurately represent the expected power of a block after it has passed through a physical design flow. We therefore chose to characterize the blocks by extracting power models after carrying them through a complete physical design flow. This process is time-consuming, but it is faster than the alternative, which is to assemble the complete system RTL before beginning a system-level study. By characterizing subblocks, we avoid the effort of designing glue logic and modifying block interfaces such that they are compatible.

Figure 4.3 illustrates our characterization flow. First, characterization parameters for each test bench are defined to link TLM simulation results to each power model.

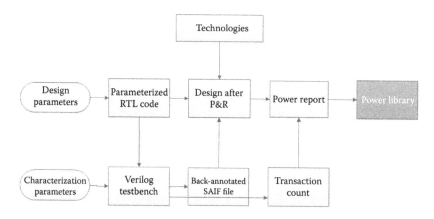

FIGURE 4.3   Flow for power model generation. (From Priyadarshi et al., In *Proceedings of IEEE International 3D System Integration Conference*, February 2012, pp. 1–8.)

An example of such a parameter is the throughput of the block (transactions per cycle), since it is related to the dynamic power consumption. Next, a standard cell netlist is generated with extracted parasitics from a commercial physical design tool. The design netlist is simulated with the characterized test bench and detailed switching activities are recorded. The power consumption is calculated using switching activities and a standard cell technology library. At last, a lookup table is generated, including block information and power consumption indexed by characterization parameters, such as the transaction rate of the block.

4.3.5 COMPOSITE MODEL EXTRACTION

The backend of the proposed flow primarily consists of composite model extraction, rough floor planning, and thermal simulation. It takes material properties and wafer and stacking technology descriptions as inputs. These data can be obtained from vendor process design kits (PDKs). In order to permit delivery of this flow and avoid the intellectual property restrictions on vendor kits, we developed a free, open-source design kit compiler for stacked dies in a predictive 45 nm technology, which we call the FreePDK3D45 [6]. The kit represents a five-tier stack of FreePDK45 predictive technology, allowing the designers to maintain a complete 3D conception of their design. Using FreePDK3D45 as an underlying framework, we have developed an open-source toolset to deliver the system-level flow presented here, called Pathfinder3D [15]. The tool currently supports power, thermal, and routability evaluations of TSV-enabled digital architectures.

Pathfinder3D's technology file format allows the specification of a new heterogeneous 3D stack with changes to only a few lines of code. The materials section allows specifying a list of material names, their thermal conductivities, densities, and specific heat capacities. Single-wafer manufacturing technologies and their associated layers, materials, and thicknesses can be specified in the wafer technology section. The stack technology section allows instantiation of wafer technology tiers plus additional glue layers. Using this approach, a complex layer stack can be described very concisely.

The complete list of cross-sectional layers generated by the FreePDK3D45 technology is more than 100, which is too much detail for system-level thermal analysis and makes it slow. Hence, the Pathfinder3D tool represents each tier in the stack with a reduced set of composite layers (composite technology file), namely, (1) substrate, (2) active, and (3) metal. This division is chosen because of the differing thermal properties of each portion. First, substrates typically have much higher conductivity than other portions and are therefore modeled separately. Second, the active layer is modeled separately, because it is typically the place where the majority of heat is generated. Third, the metallization layers can be collectively viewed as a metal-insulator composite and are typically where most of the temperature rise occurs in a 3D stack. Thermal conductivity in such composite materials can be difficult to predict since the conductivities of materials and insulators differ by a factor of 100 to 1000. Accurate prediction requires a detailed description of the structure and a set of complex matrix solutions to find the thermal conductivity (k) to be used with Fourier's law of heat conduction in the form of a 3 × 3 matrix tensor. In the absence of detailed layout information, Pathfinder3D constructs a basic unit cell that depends on the metal densities for each layer (with a range of 0 to 1) and its routing direction (horizontal, vertical, or cut). Performing the precise matrix calculation of the conductivity tensor for the basic unit cell is straightforward and described in detail in the Pathfinder3D tutorial [15]. However, the conductivity tensor is currently not computed, because of the significant implementation difficulty. Alternatively, Pathfinder3D calculates upper and lower bounds on the diagonal elements of the conductivity tensor (k_x, k_y, and k_z). The off-diagonal elements tend to be small, since most of the wires tend to align with the x, y, and, z axes.

How these conductivity bounds are calculated will now be described. The two simplest thermal conductivity calculation models are the parallel and orthogonal models. If a composite material consists of a matrix of one material with cross sections in another material running parallel to the direction of heat conduction, then the material can be modeled as a parallel combination of conductances. If a composite material consists of a cascade of material segments that are orthogonal to the direction of heat conduction, then the material can be modeled as a series combination of conductances. The parallel and orthogonal models are shown in Figure 4.4(a) and (b), respectively. Most structures will not directly correspond to either of these two categories; however, there are a large number of structures (including rectangular meshes) for which a combination of the parallel and orthogonal equivalent calculations can be applied. The order in which the calculations are performed is based on different assumptions on how heat flows and can therefore be viewed as providing bounds on the equivalent conductivity. The upper bound can be calculated by considering a unit cell consisting of N orthogonal segments, each with M_j parallel cross sections, as shown in Figure 4.5. In this case, a parallel equivalent model can be assumed for each segment, and an orthogonal equivalent model can be assumed for the segments collectively.

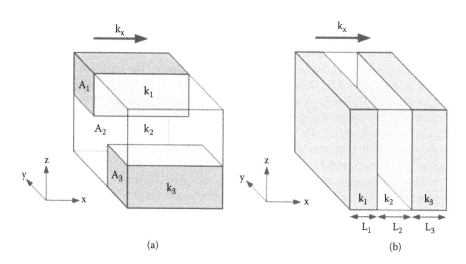

FIGURE 4.4   Thermal conductivity calculation models: (a) parallel model and (b) orthogonal model.

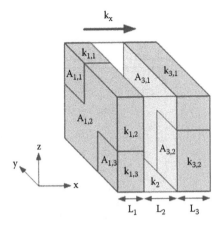

FIGURE 4.5 Cross section view of unit cell used in parallel-orthogonal conductivity calculation.

The equivalent parallel-orthogonal conductivity is calculated as

$$
k_{\text{eq\_par\_orth}} = \left[\sum_{i=1}^{N} \frac{L_i}{\left(\displaystyle\sum_{j=1}^{M_i} \frac{A_{i,j}}{K_{i,j}} \right)} \right]^{-1}
\tag{4.3}
$$

where

$$
\sum_{i=1}^{N} L_i = 1, \forall_i \sum_{j=1}^{M_i} A_{i,j} = 1
\tag{4.4}
$$

This model differs from the exact equivalent conductivity, because it assumes ideal heat spreading between the segments. In reality, heat spreads gradually through the material when a temperature gradient is applied. Therefore, $k_{\text{eq\_par\_orth}}$ can be viewed as an upper bound on the conductivity of the composite material.

The lower bound on equivalent thermal conductivity can be calculated considering a unit cell consists of N parallel cross sections, each with M_j orthogonal segments, as shown in Figure 4.6. In this case an orthogonal equivalent can be assumed for each cross section, and a parallel equivalent can be assumed for the cross sections collectively. The equivalent orthogonal-parallel conductivity is

$$
K_{\text{eq\_orth\_par}} = \sum_{i=1}^{N} \frac{A_i}{\left(\displaystyle\sum_{j=1}^{M_i} \frac{L_{i,j}}{K_{i,j}} \right)}
\tag{4.5}
$$

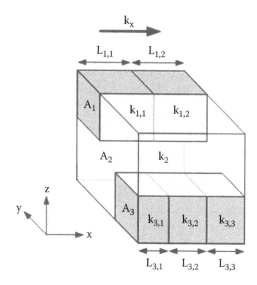

FIGURE 4.6 Cross section view of unit cell used in orthogonal-parallel conductivity calculation.

where

$$\sum_{i=1}^{N} A_i = 1, \forall_i \sum_{j=1}^{M_i} L_{i,j} = 1 \tag{4.6}$$

This model differs from the exact equivalent conductivity, because it assumes that there is no heat spreading between the cross sections. Therefore, $k_{\text{eq\_orth\_par}}$ can be viewed as a lower bound on the conductivity of the composite material.

In Pathfinder3D, the approximations leading to the upper and lower bound thermal conductivities are applied to the basic unit cell that is constructed for each technology. The upper bound, $k_{\text{eq\_par\_orth}}$, is used for k_x and k_y values, while the lower bound, $k_{\text{eq\_orth\_par}}$, is used for the k_z value. The tool also considers the effect of TSVs in the equivalent thermal conductivity calculation. TSVs are essentially cut layers, but depending on the start and stop layers defined for each TSV in the technology file, they are likely to coincide with a wafer technology layer (i.e., a horizontal or vertical routing layer or another cut layer). In these cases, the density of the TSVs is added to the density of the wafer technology layer, and the wafer technology routing direction is assumed. The new effective metal thermal conductivity is calculated as the parallel equivalent of the wafer technology routing layer and the TSV. This is akin to an upper bound on the conductivity impact of a TSV. Pathfinder3D also calculates the equivalent specific heat capacity of a metal-insulator composite using a similar approach to that used for the equivalent thermal conductivity calculation. This parameter is used in transient thermal simulation.

4.3.6   ROUGH FLOOR PLANNING

As shown in Figure 4.2, a rough floor plan and power profiles (average and transient) of floor plan blocks are required for the thermal simulations. Pathfinder3D allows a user to specify a textual description of his or her floor plan. The user can define the dimensions of basic building blocks of each tier in the 3D stack as macrocells. Each macrocell can have multiple sockets representing ports of a block. A user can then replicate these macrocells as different instances at various locations in the corresponding tiers by specifying the coordinates and the connections among instances. The instance connection information can be imported from the system description defined in the front end. A simple routing tool then can be applied to estimate the global wire length. Furthermore, the number of repeaters can be estimated using a minimum delay insertion algorithm. This allows estimating the interconnect power of the design. Pathfinder3D reads the floor plan and creates a layout using the OpenAccess database. In the layout, each macrocell is mapped with three layers: the substrate, active, and metal-insulator composite associated with the macrocell's wafer technology.

4.3.7   ELECTROTHERMAL SIMULATION

For the steady-state thermal simulation, users can specify the average power of each floor plan instance in a separate power stimulus file at the beginning of the simulation. Pathfinder3D reads the power stimulus file and assigns the power of each instance to the active layer of that instance. Currently, power is uniformly distributed across the area of a macrocell. For the dynamic electrothermal simulation, a relaxation approach is used where TLM and thermal simulators exchange power and temperature details after every certain time interval. A lock-step algorithm is used for synchronizing the electrical and thermal simulations.

The Pathfinder3D toolset consists of a physical thermal extractor, WireX [18]. WireX reads the layout generated from the rough floor plan and creates a linear resistive and capacitive thermal netlist. It uses the thermal conductivity, specific heat capacity, and other material properties of the composite model to generate this netlist. WireX meshes the layout and discretizes it in cuboids. Each cuboid is modeled using a thermal resistor from the center of the cuboid to each face, with a thermal capacitor from the center terminal to the thermal ground. The user is able to control the fidelity of the mesh. Usually for fast thermal simulation in system-level studies the resolution of the mesh is kept low. For steady-state thermal analysis, the extractor generates a thermal modified nodal admittance matrix (Y) and power vector (J). Then, $Yv = J$ is solved by a linear sparse matrix solver to get the temperature vector (v). The dynamic thermal simulation requires a transient solver. Currently, Pathfinder3D does not include any transient solver. However, it can generate a netlist compatible with HSPICE or the open-source circuit simulator fREEDA [19].

The WireX tool currently supports only one style of boundary condition, which is a perfect heat sink on one face of the stack and adiabatic on all other faces. The perfect heat sink is assumed to be connected to the first tier specified in the stack technology. All temperatures will be reported as a rise above the heat sink. For absolute temperatures, the temperature calculation must take the package and physical heat sink into account. The current model is accurate assuming the substrate of the first tier is a perfect heat spreader, which is not always an accurate assumption. However, users have flexibility to model portions of the package and heat sink as composite layers in the Pathfinder3D technology file.

4.4 CASE STUDY: QUAD-CORE 3D CHIP MULTIPROCESSOR

To demonstrate the applicability and usefulness of flow presented here, a 2D and three 3D implementations of a quad-core CMP are considered. Figure 4.7 shows the 2D floor plan (called *2D FLP*) of the quad-core CMP. Each core in the system is of dimension 1.52 × 2 mm and consumes 1.02 W of power on average and represents a four-wide out-of-order superscalar processor. Total L2 cache size is 2 MB, which is distributed in four banks as shown in Figure 4.7. The dimension and average power of each L2 bank are 2.36 × 2 mm and 100 mW, respectively. A crossbar-style router for establishing the connection between the cores and L2 cache banks is used. The dimension and power of the router are 0.5 × 0.8 mm and 20 mW, respectively.

Three 2-tier 3D implementations of the quad-core CMP are considered here: (1) *3D FLP*1, (2) *3D FLP*2, and (3) *3D FLP*3. In *3D FLP*1, at each tier there are two cores and two L2 cache banks. In this floor plan, the core on tier B is stacked over the core on tier A, where tier A is near the heat sink. Similarly, the L2 cache bank on tier B is stacked over the L2 cache bank on tier A. The router is located on tier A. TSVs are used for connecting the cores and L2 cache banks on tier B with the router on tier A. Hadlock's algorithm, a shortest-path algorithm for grid graphs [20], is used to

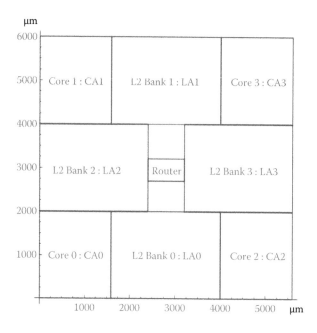

FIGURE 4.7 Floor plan of quad-core 2D CMP considered in the case study.

estimate the total global wire length, which includes the connection between router and cores, and router and cache banks. The inputs for wire length estimation process include connection coordinates, chip area, blockage layers, TSV size and spacing, number of metal layers used in wiring, wire width and spacing, and resolution of the routing grid. Hadlock's algorithm divides the floor plan into small grids, which includes the capacity of vertical/horizontal wires, maximum number of TSVs allowed, and block layer information. Using the coordinates of wire endpoints, the algorithm finds the shortest path for each wire. Once a wire is successfully routed, the wire capacities and TSV information are updated, which can affect the routing result for following wires to avoid overcongestion scenarios. After the routing, total global wire length, TSV distribution, wire length distribution, and wire power consumption are calculated.

3D FLP2 is similar to 3D FLP1 except in this case, cores on tier B are stacked over L2 cache banks on tier A, and L2 cache banks on tier B are stacked over cores on tier A. In 3D FLP3, all four cores and the router are located on tier A. Tier B has all four L2 cache banks. This represents a static random access memory (SRAM) over logic case.

Figure 4.8 shows the spatial thermal profile of the floor plan shown in Figure 4.7, which corresponds to 2D implementation of quad-core CMP. This profile is obtained using static thermal simulation and shows the rise in temperature (in Kelvin) and spatial thermal gradient. The hotspots are located on the corners because of high power density in core regions. L2 regions are relatively cooler. The maximum spatial temperature gradient across a grid block is 4.02 K.

Figure 4.9 shows the spatial thermal profile corresponding to tier B (away from the heat sink) of floor plan 3D FLP1. The temperature rise in this case is significantly more than 2D implementation because of increased power density in the 3D stack.

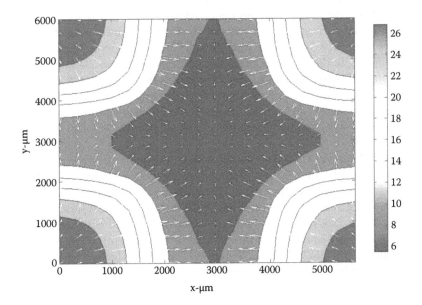

FIGURE 4.8   Spatial thermal profile corresponding to 2D implementation of quad-core CMP.

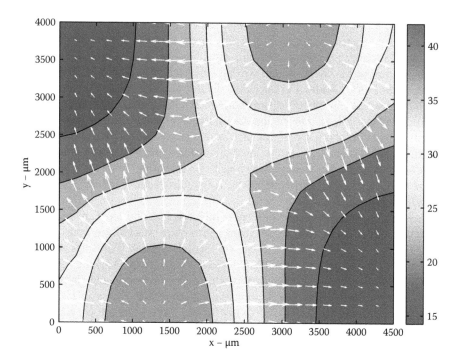

FIGURE 4.9   Spatial thermal profile corresponding to tier B of floor plan *3D FLP*1.

The hotspots are located in two regions where cores are stacked on top of each other. In this case, the maximum spatial temperature gradient across a grid block is 4.51 K. In the thermal simulations presented here, a heat sink-based cooling solution with forced-air convection is considered. A single lumped thermal convection resistance of 0.8 K/W is used. Figure 4.10 shows the spatial thermal gradient of floor plan *3D FLP*2. In this case, the temperature rise is more than 2D implementation but less than 3D implementation, *3D FLP*1. This is because in *3D FLP*1 two more heat dissipating blocks, i.e., cores, are stacked on top of each other, whereas in *3D FLP*2, a core is stacked on top of a L2 cache bank, which dissipates much less power than a core. In this case, the maximum spatial temperature gradient across a grid block is 2.19 K.

Figure 4.11(a) and (b) shows the floor plan and spatial thermal profile of *3D FLP*3, respectively. In this case, the temperature rise is slightly more than 2D implementation but lower than other 3D implementations. This is because major heat dissipating sources (i.e., cores) are on the tier near the heat sink (i.e., tier A), allowing better cooling efficiency. The hotspots are located in the regions where two cores are situated nearby. In this case, the maximum spatial temperature gradient across a grid block is 1.26 K. Figure 4.12 shows the dynamic thermal profile of 2D and 3D floor plans. A transient power trace is required for the dynamic simulation, which is obtained by executing *gcc* workload from SPEC CPU2000 on cores on tier A and *bzip* workload from SPEC CPU2000 on cores on tier B on a TLM simulator. Figure 4.12 also illustrates floor plan *3D FLP*3 is thermally better

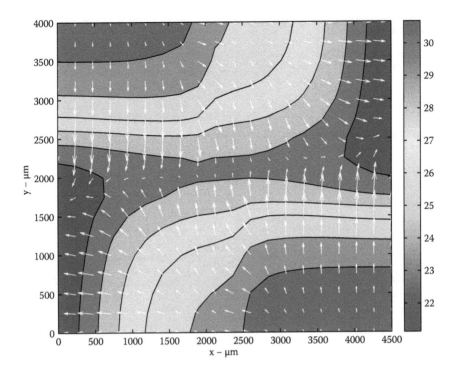

FIGURE 4.10   Spatial thermal profile corresponding to tier B of floor plan *3D FLP2*.

than the other 3D floor plans, *3D FLP1* and *3D FLP2*. Furthermore, *3D FLP1* is significantly hotter than other floor plans.

Table 4.2 presents a summary of results obtained by comparing 2D and 3D implementations in terms of total global wire length, maximum temperature rise, and maximum spatial temperature gradient across a grid block. Note that here total global wire length corresponds to sum of length of wire required to connect all the cores and L2 cache banks to a router. All three 3D floor plans have total global wire length smaller than that of the 2D floor plan. The results in the table show that 3D integration can significantly reduce the global wire length, which can reduce the interconnect delay and power. The floor plan *3D FLP3* facilitated more reduction in global wire length than that of floor plans *3D FLP1* and *3D FLP2*. This is due to an uneven area of tiers A and B in the case of *3D FLP3*. In floor plan *3D FLP3*, all the L2 cache banks are on tier B; thus, it is bigger in area than tier A (see Figure 4.11(a)). Removing the L2 banks from tier A provides more spacing, and thus reduces the congestion for wires connecting to the router. The areas of tiers A and B in the case of floor plans *3D FLP1* and *3D FLP2* are the same because each tier contains two cores and two L2 cache banks. However, the presence of L2 cache banks on tier A blocks several routing tracks, and thus increases the routing congestion, resulting in larger wire length. This case study shows that using the flow presented in Section 4.3, comparisons like those shown in Table 4.2 can be quickly performed at the system level without any detailed implementation, guiding system architects in eliminating bad design choices early on.

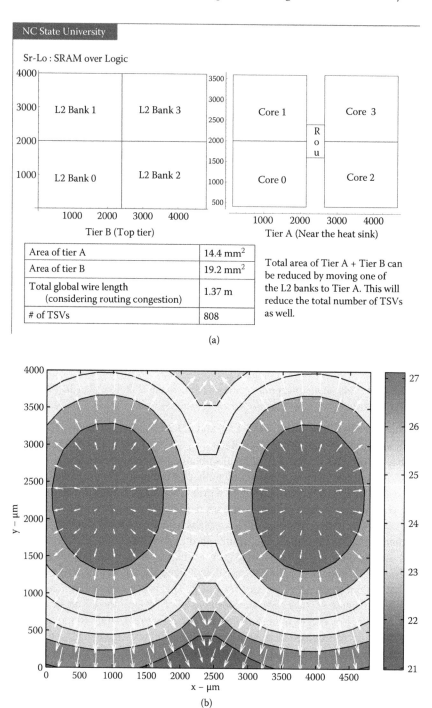

FIGURE 4.11   (a) Floor plan of *3D FLP3*. (b) Spatial thermal profile corresponding to tier B of *3D FLP3*.

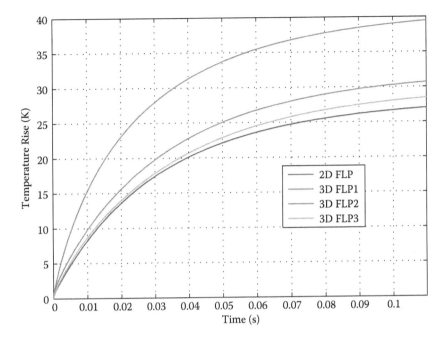

FIGURE 4.12   Dynamic thermal profile of 2D and 3D floor plans of quad-core CMP.

TABLE 4.2
Comparison of 2D and 3D Floor Plans

Metric	2D FLP	3D FLP1	3D FLP2	3D FLP3
Total global wire length	3.016 m	2.22 m	1.71 m	1.37 m
Maximum rise in temperature	26.2 K	39.8 K	30.5 K	28.25 K
Maximum spatial temperature gradient	4.02 K	4.51 K	2.19 K	1.26 K

4.5   CONCLUSION

This chapter discussed the thermal challenges of 3D ICs and necessity of electro-thermal simulation. A system-level electrothermal simulation flow for 3D ICs is presented, which allows us to study the trade-offs between various design choices for 3D integration early in the design phase. In the system-level design space exploration trends are more important than the actual values, and traditional flows typically ignore the physical details in such explorations. However, not considering the physical details can skew the trends. This effect is more pronounced in the case of 3D ICs because 3D manufacturing options can significantly impact the system-level design choices. The flow presented here brings an appropriate level of 3D IC specific physical awareness to system-level flows. The applicability of the flow is shown using a case study comparing the thermal characteristics of a 2D and three different 3D implementations of a quad-core chip multiprocessor.

REFERENCES

1. Davis, W. R., Wilson, J., Mick, S., Xu, J., Hua, H., Mineo, C., Sule, A., Steer, M. B., and Franzon, P. D., Demystifying 3D ICs: the pros and cons of going vertical, *IEEE Design and Test of Computers*, 2005, 22(6), 498–510.

2. Banerjee, K., Souri, S., Kapur, P., and Saraswat, K., 3-D ICs: a novel chip design for improving deep-submicrometer interconnect performance and systems-on-chip integration, *Proceedings of the IEEE*, 2001, 89(5), 602–633.

3. Bernstein, K., Andry, P., Cann, J., Emma, P., Greenberg, D., Haensch, W., Ignatowski, M., Koester, S., Magerlein, J., Puri, R., and Young, A., Interconnects in the third dimension: design challenges for 3D ICs, In *Proceedings of IEEE/ACM Design Automation Conference*, June 2007, pp. 562–567.

4. Franzon, P. D., Davis, W. R., Steer, M. B., Lipa, S., Oh, E., Thorolfsson, T., Melamed, S., Luniya, S., Doxsee, T., Berkeley, S., Shani, B., and Obermiller, K., Design and CAD for 3D integrated circuits, In *Proceedings of IEEE/ACM Design Automation Conference*, June 2008, pp. 668–673.

5. Priyadarshi, S., Choudhary, N., Dwiel, B., Upreti, A., Rotenberg, E., Davis, W. R., and Franzon, P. D., Hetero² 3D integration: a scheme for optimizing efficiency/cost of chip multiprocessors, presented at International Symposium on Quality Electronic Design, March 2013.

6. The FreePDK3D45 Predictive 3D IC Process Design Kit, http://www.eda. ncsu.edu/wiki/FreePDK3D45:Contents.

7. Higher abstraction saves more power, http://www.iqmagazineonline.com/current/pdf/Pg30–33_IQ_32–Higher_Abstraction_Saves_More_Power.pdf.

8. Wunsche, S., Clauss, C., Schwarz, P., and Winkler, F., Electro-thermal circuit simulation using simulator coupling, *IEEE Transactions on Very Large Scale Integration Systems*, 1997, 5(3), 277–282.

9. Petegem, W., Geeraerts, B., Sansen, W., and Graindourze, B., Electrothermal simulation and design of integrated circuits, *IEEE Journal of Solid-State Circuits*, 1994, 29(2), 143–146.

10. Chiang, T. Y., Banerjee, K., and Saraswat, K., Compact modelling and SPICE-based simulation for electrothermal analysis of multilevel ULSI interconnects, In *Proceedings of IEEE/ACM International Conference on Computer Aided Design*, January 2001, pp. 165–172.

11. Szekely, V., Poppe, A., Rencz, M., Csendes, A., and Pahi, A., Electro-thermal simulation: a realization by simultaneous iteration, *Microelectronics Journal*, 1997, 247–262.

12. Digele, G., Lindenkreuz, S., and Kasper, E., Fully coupled dynamic electro-thermal simulation, *IEEE Transactions on Very Large Scale Integration Systems*, 1997, 5(3), 250–257.

13. Rencz, M., Szekely, V., Poppe, A., and Courtois, B., Algorithmic and modelling aspects in the electro-thermal simulation of thermally operated Microsystems, In *Proceedings of 6th International Conference on Modelling and Simulation of Microsystems*, 2003, pp. 476–479.

14. Priyadarshi, S., Hu, J., Choi, W. H., Melamed, S., Chen, X., Davis, W. R., and Franzon, P. D., Pathfinder 3D: a flow for system level design space exploration, In *Proceedings of IEEE International 3D System Integration Conference*, February 2012, pp. 1–8.

15. The Pathfinder3D 3DIC architecture evaluator tool, http://research.ece.ncsu.edu/pathfinder.

16. Beltrame, G., Sciuto, D., and Silvano, C., Multi-accuracy power and performance transaction-level modeling, *IEEE Transactions on Computer-Aided Design of Integrated Circuits and System*, 2007, 26(10), 1830–1842.

17. Martin, M. M. K., Sorin, D. J., Beckmann, B. M., Marty, M. R., Xu, M., Alameldeen, A. R., Moore, K. E., Hill, M. D., and Wood, D. A., Multifacet's general execution-driven multiprocessor simulator (GEMS) toolset, *ACM SIGARCH Computer Architecture News*, 2005, 33(4), 92–99.

18. Melamed, S., Thorolfsson, T., Harris, T. R., Priyadarshi, S., Franzon, P. D., Steer, M. B., and Davis, W. R., Junction-level thermal analysis of 3-D integrated circuits using high definition power blurring, *IEEE Transactions on Computer-Aided Design of Integrated Circuits and Systems*, 2012, 31(5), 676–689.

19. fREEDA, an open-source multi-physics simulator, http://www.freeda.org/.

20. Hadlock, F. O., A shortest path algorithm for grid graphs, *Networks*, 1977, 7(4), 323–334.

5 Thermal Management for 3D ICs/Systems

Francesco Zanini

CONTENTS

5.1   INTRODUCTION

5.1.1   MOTIVATION

Energy consumption has become one of the primary concerns in electronic design. As can be noted, Figure 5.1 shows how the thermal profile of a commercial *multi-processor system-on-chip* (MPSoC) is quite nonuniform and hot spots (localized MPSoC areas with unsafe working temperature) may arise. This effect not only affects the performance of the system, but also leads to unreliable circuit operation and affects the lifetime of the chip [38]. Thus, thermal management for multicore architectures is a critical matter to tackle.

In the last years, thermal management and balancing techniques received a lot of attention. Many state-of-the-art thermal control policies operate power management

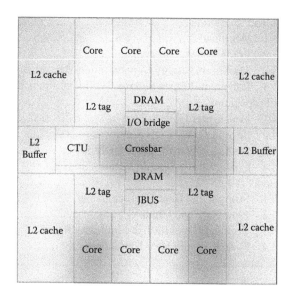

FIGURE 5.1   Thermal profile of Sun UltraSPARC T1 (Niagara) platform. (From Kongetira et al., Niagara: A 32-Way Multithreaded SPARC Processor, In *IEEE MICRO*, 2005.)

by employing *dynamic frequency and voltage scaling* (DVFS)-based techniques [15, 26]. This technique scales down the frequencies and the voltages of some specified units to save power and optimize performance. The problem with this technique is that the frequent abrupt change in working frequencies and voltages produces thermal cycling, which raises the failure rate of the system [14, 45]. In addition, discontinuous power mode transitions, in both voltage and frequencies scaling, waste additional power [18]. For the aforementioned reasons there are many trade-offs in power management techniques that are not easy to handle properly during the run-time execution of the MPSoC and with a low computational overhead.

Moreover, new challenges are related to emerging technologies for heat extraction. Heat extraction is based on a heat sink. A heat sink is a component or assembly that transfers heat generated within a solid material to a fluid medium, such as air or a liquid. Examples of heat sinks are the heat exchangers used in refrigeration and air conditioning systems and the copper dissipator placed on top of microprocessors in desktop computers. A heat sink is physically designed to increase the surface area in contact with the cooling fluid surrounding it, such as the air. In past decades the research to increase the heat extraction performed by this element was focused on changing design factors such as air velocity, choice of material, fin (or other protrusion) design, and surface treatment. In recent years, with the increase of power density, the research has been focused on changing the coolant fluid. Not only air is used nowadays, but also fluids circulating and exchanging heat with the heat spreader. The pipes where the coolant liquid is circulating can be either integrated inside MPSoCs [22, 23] or placed on top of them [35–37], as shown in Figure 5.2. These new cooling technologies represent a key perspective for novel thermal management policies to both reduce cooling power consumption and at the same time increase performance.

5.1.2 DYNAMIC VOLTAGE AND FREQUENCY SCALING

Dynamic voltage and frequency scaling (DVFS) is a technique to reduce energy consumption by changing processor speed and voltage at runtime, depending on the needs of the applications running. This method is widely used as part of strategies to manage switching power consumption in battery-powered devices such as cell phones and laptop computers. Low-voltage modes are used in conjunction with lowered clock frequencies to minimize power consumption associated with components

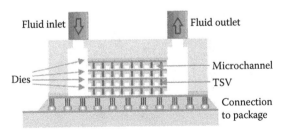

FIGURE 5.2 Cross section of a test stack with liquid cooling.

such as central processing units (CPUs) and digital signal processors (DSPs); only when significant computational power is needed will the voltage and frequency be raised. Special supply regulation circuits are required to be able to deliver a multiple set of voltages to the SoC.

It is important to vary both the voltage and the frequency because if only processor frequency is scaled, the total energy savings would be small or zero, as power is inversely proportional to cycle time and energy is proportional to the execution time and power. The switching power dissipated by a chip using static complementary metal oxide semiconductor (CMOS) gates is CV^2f, where C is the capacitance being switched per clock cycle, V is voltage, and f is the switching frequency, so this part of the power consumption decreases quadratically with voltage [32]. The formula is not exact, however, as many modern chips are not implemented using only CMOS, but also use pseudo-nMOS gates, domino logic, etc.

Moreover, there is also a static leakage current, which has become more and more accentuated as feature sizes have become smaller (below 90 nm) and threshold levels lower. When leakage current is a significant factor in terms of power consumption, chips are often designed so that portions of them can be powered completely off. This capability is an additional challenge to be managed in new thermal management systems.

5.1.3   3D-MPSoC AND LIQUID COOLING

A *three-dimensional integrated circuit* (3D IC) is a chip in which two or more layers of active electronic components are integrated both vertically and horizontally into a single circuit. 3D integration [2] is a recently proposed design method for overcoming the limitations with respect to delay, bandwidth, and power consumption of the interconnects in large multiprocessor system-on-chip (MPSoC) chips, while reducing the chip footprint and improving the fabrication yield. The main reason of all these benefits is the introduction of connections from one die to the other. These vertical wires are called *through-silicon vias* (TSVs), and they allow us to make connections shorter than in normal 2D chips [25]. A simplified illustration of a TSV in a 3D stack is shown in Figure 5.3.

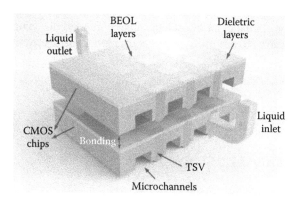

FIGURE 5.3   Simplified illustration of 3D stack with intertier liquid cooling.

Figure 5.3 is an example of 3D-MPSoC where liquid cooling (liquid inlet/liquid outlet) is used as the cooling mechanism. The reason for using liquid cooling is because of the higher thermal resistivity [16, 31], which irregularly spreads in the 3D chip stack. Hence, it is more difficult to remove the heat from 3D systems with respect to conventional 2D-MPSoCs.

Conventional back-side heat removal strategies, such as air-cooled heat sinks and microchannel cold plates, only scale with the die size and are insufficient to cool 3D-MPSoC with hot spot heat fluxes up to 250 W/cm^2, as expected in forthcoming 3D-MPSoC stacks [6]. On the contrary, intertier single- and two-phase liquid cooling are potential solutions to address the high temperatures in 3D-MPSoCs, due to the higher heat removal capability of liquids in comparison to air [8].

The use of convection in microchannels to cool down high-power-density chips has been an active area of research since the initial work by Tuckerman and Pease [44]. The heat removal capability of interlayer heat transfer with pin-fin in-line structures for 3D chips is investigated in Brunschwiler et al. [6].

5.1.4 THERMAL MODELING

Skadron et al. [40] and Paci et al. [29] have developed a thermal power model for super-scalar architectures. It not only predicts the temperature variations between the different components of a processor, but also accounts for the increased leakage power and reduced performance. Their results clearly prove the importance of hot spots in high-performance systems. Based on this model, many architectural extensions have been proposed.

The model exploits the well-known analogy between electrical circuits and thermal models. It decomposes the silicon die and heat spreader in elementary cells that have a cubic shape (Figure 5.4) and use an equivalent RC model for computing the temperature of each cell. By varying the cell size, we can trade off the simulation speed of the thermal model with its accuracy. The coarser the cells become, the fewer cells we need to simulate, but the less accurate the temperature estimates become.

A thermal capacitance and five thermal resistances are associated with each cell (Figure 5.5). Four resistances are used for modeling the horizontal thermal spreading, whereas the fifth is used for the vertical thermal behavior. See [29] for further details.

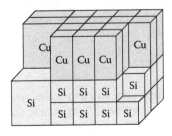

FIGURE 5.4 Cubic cells model of the MPSoC.

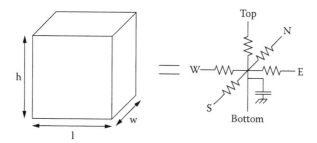

FIGURE 5.5   Equivalent RC circuit of a cell.

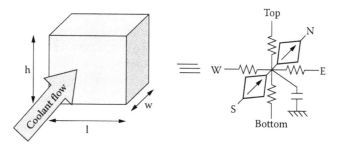

FIGURE 5.6   Equivalent circuit of a fluid thermal cell.

These concepts model with good accuracy 2D and 3D chips using air cooling. In the case of liquid cooling, the model has been modified to the one presented by Sridhar et al. [41]. This model is exactly the same for any silicon or metal structure. The only difference is the introduction of the way the liquid cooling flow has been modeled. A graphical representation is shown in Figure 5.6.

Cells corresponding to microchannels containing the cooling fluid are modeled like any other cell with the addition of a current source and a current sink. To model the heterogeneous characteristics of the variable flow rate in microchannels, Sridhar et al. [41] introduced the ability to change the intensity of the current source and the resistance value of the cell at runtime. Thus, the interlayer material composing the MPSoC is divided into grid cells, where each grid cell except for the cells of the microchannels has a fixed thermal resistance value, depending on the characteristics of the interface material. Thermal properties of the microchannel cells are computed based on the liquid flow rate through the cell at runtime as presented in Sridhar et al. [41].

5.1.5   THERMAL SENSING

Any thermal management algorithm to control the system must know the overall state (or thermal profile) of the MPSoC. This means that the temperature of every single cell in which the floor plan has been divided must be known.

A study of the thermal profile estimation problem has been presented in Memik et al. [24] and Sharifi and Rosing [39]. The proposed solutions are based on techniques trying to reduce temperature differences between thermal sensors and hot spots by using the minimum possible number of sensors for a certain accuracy. The problem with these approaches is that since hot spots are application dependent, there is no guarantee that all hot spots are detected during the lifetime of the device.

A better approach to estimate these temperatures is to use a state estimator [13]. A state or thermal profile estimator is an algorithm able to derive the current thermal profile based on measurements in some specific locations on the chip with a specific rate. The parameter that measures how much a system is observable is called observability. Observability refers to the property of a system that enables the reconstruction of the state variables given the inputs [13]. It means that we are able to reconstruct completely the thermal profile of the chip given the inputs only by looking at the measurements coming from the sensors.

The placement problem in an MPSoC is the problem of selecting the right locations of thermal sensors to both minimize the number of sensors and maximize the observability of the system. Sumana and Venkateswarlu [43] select the location of the sensing element according to a sensor strategy using the just described observability concept. Joshi and Boyd [17] solved the problem of making a system observable by employing of graph theory. The problem of choosing a set of measurements from a much larger set that also minimizes the estimation error is solved by Boukhobza and Hamelin [4] using a convex optimization-based approach. This last method approximately solves the problem and has no guarantee that the performance gap is always small.

5.2 MATHEMATICAL MODELS

5.2.1 SYSTEM ENERGY

Saving power, improving performance, and increasing chip reliability are three important challenges facing MPSoC designers. These three objectives must be reconciled with the fact that the MPSoC has to execute all the workload requested from the scheduler with the minimum power consumption and the smallest acceptable latency.

Figure 5.7 shows the status of the system from the scheduler point of view at a certain fixed time point k. The frequency of the sampling point is the frequency with which thermal policies are applied. The term R stands for tasks that are still in the processor and need to be completed, W are tasks arrived, waiting to be executed, X_i is the task workload arrived i previous time steps, N is the length of the past task arrival history, and D is the task workload that will arrive in the next time step between the actual one, k, and the next one, $k + 1$.

Since the task arrival is a stochastic process, a prediction must be made in relation to the workload that will be required by the scheduler in the next policy observation period D. If the scheduler employs a dynamic scheduling in the time frame

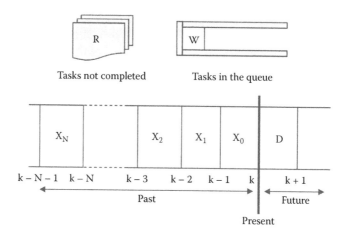

FIGURE 5.7   Scheduler viewpoint snapshot of the MPSoC at time k.

corresponding to a window, it is not possible to know the workload in time frame D at time k in advance.

The ideal solution would be a method able to exactly estimate the number of tasks arriving in the next time slot. The better the estimation is, the better the performance of the thermal management system will be.

5.2.1.1   Energy Efficiency Quantification

We can assume a linear relation between the frequency of operation of a core and the amount of instructions it executes [10]. We consider now two consecutive window frames A and B, and we define as f_A and f_B the frequencies that will fulfill the scheduler requirements in the two consecutive time windows A and B. The picture in the top of Figure 5.8 shows the setup. As can be noted, the queue is empty since all tasks Ta and Tb are executed, respectively, in windows A and B.

To compute the power consumption in this ideal case we use derivations from Rao et al. [34]. In his work, he showed that the relation between the frequency and the power can be approximated as a polynomial function with a degree larger than 1. The formula is quite complex. A good approximation is represented by the following equation:

$$P(f_A) = K_p \cdot f_A^\alpha + L \tag{5.1}$$

where f_A is a frequency value, K_p, α, and L are constants, and $P()$ is the function that relates the frequency of a functional unit to its power dissipation. The term $K_p \cdot f_A^\alpha$ models the active power consumption. The constant term L models the part of the power that does not depend on the frequency, such as the leakage power consumption.

Using Equation 5.1, the power consumption P_{ideal} due to the execution of tasks Ta and Tb in this scenario is given by the following equation:

$$P_{ideal} = K_P \cdot f_A^\alpha + K_P \cdot f_B^\alpha + 2L \tag{5.2}$$

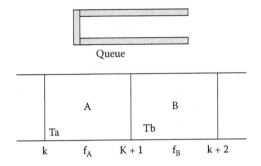

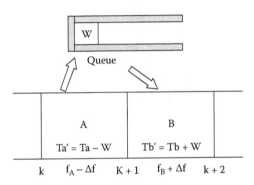

FIGURE 5.8 Effect of a frequency estimation error from a frequency perspective in a DVFS system. Top: Ideal case. Bottom: Real case with an estimation error Δf.

The bottom of Figure 5.8 shows a scenario where an estimation error Δf has occurred. Because the frequency is lower than expected, some tasks (W) are not executed and are stored in the queue at time step $k + 1$. To compare both scenarios of Figure 5.8, we must have the same starting and ending conditions. This implies having all jobs executed by the time step $k + 2$. To achieve that, queuing tasks need to be executed in time frame B. Thus, a frequency higher than the ideal one has to be used. The power consumption in this scenario P_{err} due to the execution of tasks Ta' and Tb', respectively, in time frames A and B is given by the following equation:

$$P_{err} = K_P \cdot \left[(f_A - \Delta f)^\alpha + (f_B + \Delta f)^\alpha \right] + 2L \qquad (5.3)$$

We define the energy loss E due to an estimation error as

$$E = P_{err} - P_{ideal} \qquad (5.4)$$

By substituting Equations 5.2 and 5.3 into Equation 5.4, we obtain

$$E = K_P \cdot \left[(f_A - \Delta f)^\alpha + (f_B + \Delta f)^\alpha - f_A^\alpha - f_B^\alpha \right] \qquad (5.5)$$

If α is greater than 1, the energy loss E in Equation 5.5 is greater than zero.

As a concluding remark, if there is a frequency estimation error, some of the tasks will stay in the queue and will be executed in the next time slot. This will cause the processor to run at a higher frequency in the next slot since now there are some extra jobs pending due to the estimation error. If the relation between the frequency and the power consumption were a linear function, there would be no waste of energy. Since the function is convex, Equation 5.5 holds, and so an estimation error is crucial from a power-performance efficiency perspective.

5.2.1.2   Energy Bounds

In Section 5.2.1.1 we analyzed the effect of a workload estimation error on the power consumption needed to execute a specific workload. In quantifying the energy efficiency, we assumed that tasks Ta and Tb are executed, respectively, in time windows A and B. This constraint makes the task delay equal to zero. Nevertheless, it is not the minimum value of power consumption that we can get to execute all tasks $Ta + Tb$ during time windows A and B.

The empirical law that represents the power consumption as a function of the frequency setting expressed by Equation 5.1 is a convex function [5]. By applying basic properties of convex functions, we obtain the following:

$$\mathbf{p}_\tau + \mathbf{p}_{\tau+\varepsilon} \geq 2 \cdot \mathbf{p}_{(\tau+\varepsilon)/2} \quad \forall \varepsilon \in [0,1) \tag{5.6}$$

where $\mathbf{p}_\tau$ is the power consumption at time τ. Since frequency setting and executed workload are positively correlated (see [10]), energy savings demand a uniformly distributed workload. The problem is that workloads are usually not uniformly distributed during the runtime execution of the policy, and scheduling task uniformly would increase latency.

Inequality 5.6 expresses this issue as follows:

$$\mathbf{p}_{pow} \leq \mathbf{p} \leq \mathbf{p}_{perf} \tag{5.7}$$

by indicating that power consumption $\mathbf{p}$ is bounded between two values. On the one hand, the lower bound ($\mathbf{p}_{pow}$) is the power value consumed when the workload is uniformly distributed. In this case, we optimize the execution for power minimization by allowing a nonzero task execution delay, but at the same time, we require that the complete workload has to be executed.

On the other hand, the upper bound ($\mathbf{p}_{perf}$) is the power consumed when all tasks are executed at the same time they arrive. In this case, we optimize the execution for performance and the resulting task execution delay is zero. Clearly, the gap between these two numbers is highly dependent on the workload properties. Figure 5.9 shows an example of the resulting power consumption versus delayed workload for different optimization criteria ranging from power-oriented to performance-oriented optimizations.

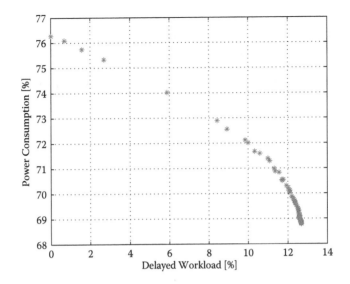

FIGURE 5.9 Example of normalized power consumption versus delayed workload for different optimization criteria ranging from power-oriented to performance-oriented optimizations.

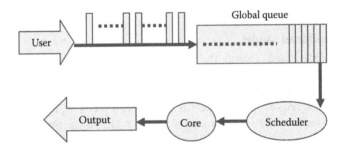

FIGURE 5.10 Overview of the system architecture.

5.2.2 WORKLOAD MODEL

5.2.2.1 System Architecture

From an operating system perspective, the interaction of the MPSoC with a user generates a task arrival process. To simulate this interaction, we use benchmarks. The benchmark is a set of programs, or other operations, to assess the relative performance of the MPSoC. Figure 5.10 shows an overview of the MPSoC system architecture abstraction.

Incoming tasks are first stored in a global queue. Each task has a specific execution time. In each scheduling cycle, the scheduler fetches the next available packet from the global queue and then dispatches the packet into a specific core for task processing. The result is the desired output.

5.2.2.2 Task Arrival Process

The task arrival is a stochastic process, as described in Figure 5.11. The picture shows the status of the system at a certain point k. The system is discrete-time, and it is sampled every T_p, where T_p is the period of time that passes between the sample at time k and the sample at time $k + 1$. The frequency of the sampling points is the frequency with which thermal policies are applied. The term R stands for the amount of work to be executed by tasks that are still in the processor and need to be completed. W is the amount of work to be executed by tasks that arrive waiting to be processed. X_i is the amount of work executed by tasks that arrive i time steps before the current step K. N is the length of the task arrival history. D is the amount of work to be executed by tasks that will arrive in the next time step between the actual one, k, and the next one, $k + 1$.

In the block diagram of Figure 5.11, the overall amount of work S to be executed between time K and time $K + 1$ equals:

$$S = R + W + D \tag{5.8}$$

Equation 5.8 expresses that the system has to execute all uncompleted tasks (R) plus the ones not executed in previous time steps that are queuing (W) plus the one that will arrive in the next time step (D). If the overall amount of tasks S are executed, performance requirements are satisfied; otherwise, the system will experience a performance loss.

5.2.2.3 Workload Model

For each p clock islands (cores), the workload is defined as the minimum value of the clock frequency that the functional unit should have to execute the required tasks within the specified system constraints.

The workload requirement at time τ is defined as a vector $\mathbf{w}_\tau \,\varepsilon\, \Re^p$, where $(\mathbf{w}_\tau)_i$ is the workload requirement value for input i at time τ. $(\mathbf{w}_\tau)_i$ is the frequency that cores

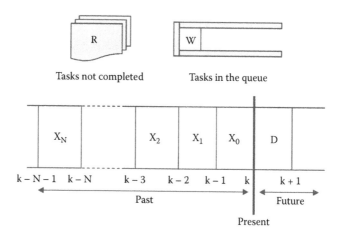

FIGURE 5.11   Snapshot of the task arrival process at time k.

associated with input i from time τ to time $\tau + 1$ should have in order to satisfy the desired performance requirement coming from the scheduler. The workload requirement is related to the overall amount of work to be executed, S, by the following equation:

$$S = T_p \cdot \sum_{i=1}^{p} (\mathbf{w}_\tau)_i \tag{5.9}$$

where T_p is the period between time k and $k + 1$ and p is the number of cores. Equation 5.9 expresses the fact that the sum of all the workloads integrated over the time period T_p should be equal to S, the overall amount of work to be executed.

Our model is assumed to be continuous and ranging from $\mathbf{f}_{min}$ to $\mathbf{f}_{max}$, the maximum frequency at which the cores can process data, namely,

$$\mathbf{f}_{min} \leq \mathbf{w}_\tau \leq \mathbf{f}_{max} \ \forall \tau \tag{5.10}$$

When $(\mathbf{w}_\tau)_i > (\mathbf{f}_\tau)_i$, the workload cannot be processed, and so it needs to be stored (W) and rescheduled in the following clock cycles. The way we measure the performance of the system in achieving the requested workload requirements at time τ is given by the vector $\mathbf{u}_\tau \ \varepsilon \ \Re^p$.

$$\mathbf{u}_\tau = \mathbf{w}_\tau - \mathbf{f}_\tau \tag{5.11}$$

We call $\mathbf{u}_\tau$ the undone workload at time τ, and it expresses the difference at time τ between the requested workload and the workload that is actually executed by the MPSoC.

5.2.2.4 Frequency and Power Model

We model the MPSoCs as synchronous with p clocks that are viewed as the inputs to the system: vector $\mathbf{f}_\tau \ \varepsilon \ \Re^p$ represents the value of the clock frequencies at time τ. The frequency value of input i at time τ is $(\mathbf{f}_\tau)_i$. Clock frequencies are continuous and range from zero to a max frequency value f_{max}. The previous statement is expressed by Inequality 5.12.

$$0 \leq \mathbf{f}_\tau \leq \mathbf{f}_{max} \ \forall \tau \tag{5.12}$$

where the symbol $\leq$ means element-wise comparison, $f_{max} \cdot \mathbf{1} = \mathbf{f}_{max}$, and $\mathbf{1}$ is a vector of all ones of size p. The frequency vector represents our optimization variable.

At time τ, the relation between the normalized value of power dissipation $\mathbf{p}_\tau \ \varepsilon \ \Re^p$ and the normalized frequency of operation $\mathbf{f}_\tau$ is expressed by Equation 5.13.

$$\mu \mathbf{f}_\tau^\alpha = \mathbf{p}_\tau \ \forall \tau \tag{5.13}$$

where μ is a technology-dependent coefficient. The constant α depends on the technology as well, and usually it takes a value between 1 and 2. If $\alpha = 1$, we have a linear dependence (i.e., frequency scaling), while if $1 < \alpha \leq 2$, we obtain a quadratic or subquadratic dependence (i.e., DVFS) [27].

We calculate the leakage power of processing units inside the MPSoC as a function of their area and actual runtime temperature. We use a base leakage power density of 0.25 Wmm^2 at 383 K for 90 nm technology according to experimental results from Bose [3]. Thus, the leakage power at a temperature T (K) is given by

$$P(T) = P_o \cdot e^{\beta(T-383)} \qquad (5.14)$$

where P_o is the leakage power at 383 K, and β is a technology-dependent coefficient. We set $\beta = 0.017$ according to experimental results from Sabry et al. [36].

5.2.3   WORKLOAD PREDICTION

5.2.3.1   Workload Arrival Process

The workload arrival process can be modeled as a stochastic process. Without loss of generality, we consider an MPSoC with eight cores. The graph in Figure 5.12 shows the duty cycle mean and standard deviation of cores in the case of light and heavy scenarios.

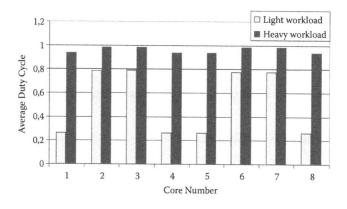

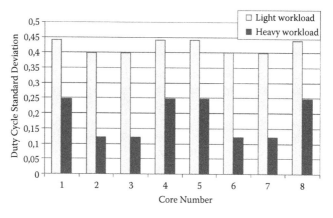

FIGURE 5.12   Duty factor mean and standard deviation of cores during system operation.

In both cases the standard deviation of the duty cycle is not zero. The reason is because the overall task execution process can be seen as a stochastic **G/G/8** process with an infinite buffer length. The first two **G**s mean that both the task arrival and the task execution time can have any distribution, and **8** is the number of cores processing the tasks. From queuing theory we know that in these systems core utilization has a standard deviation that is higher when the core utilization is lower. This is because the probability that the queue is empty is higher, and if the queue is empty, some processors may stay idle while others are busy.

Figure 5.12 shows that, in case of a light workload, the standard deviation of core duty cycle is higher than 40%, while in the other case, it is lower than 25%. Moreover, the standard deviation assumes different values on different cores. Looking at the mean value of core duty cycle, the assumption of having a core utilization that is equal to 1 and the same for every core can be seen as a 10% approximation in heavy workload scenarios, but does not hold in light ones.

In the next subsections, we propose two estimators to predict future workload requirements in time-varying scheduling scenarios employing dynamic scheduling techniques.

5.2.3.2 Prediction Accuracy

It is possible to take into account the accuracy of the prediction in the problem formulation of a thermal management policy by embedding a weighting vector γ_t. It is defined according to Equation 5.15:

$$\gamma_t = \beta_t - \left\| \mathbf{w}_t - \hat{\mathbf{w}}_t \right\| \forall \ N-L \leq t \leq N \tag{5.15}$$

where $\hat{\mathbf{w}}_t \ \varepsilon \ \Re^p$ is the workload predicted at time t by the workload predictor, $\mathbf{w}_t \ \varepsilon \ \Re^p$ is the actual value of it, and p is the number of different workloads requested at a specific time t. They correspond also to the number of processing cores of the MPSoC. The absolute value of the difference between the two represents the prediction error.

$\beta_t \ \varepsilon \ \Re^p$ is a vector that adds a penalty for the workload that has been predicted, but not executed yet, in different and future time frames. This penalty function β_t can be chosen to be linear, quadratic, exponential, or in any other way, according to the impact that a delayed execution of tasks has on performance. Indeed, the more reliable the prediction is, the smaller the prediction error is and so the bigger γ_t is. This means that since in our formulation the prediction is reliable, importance is given to the cost function corresponding to that future time frame.

5.2.3.3 Maximum Energy Concentration-Based Estimator

This system is based on both a queue and a mean value prediction based on task arrival history. The derivation presented here is referred to Figure 5.7. The proposed frequency f_p to execute all the tasks by the time $K + 1$ will be

$$f_p = \frac{R+W+D}{p \cdot T_p} \tag{5.16}$$

where T_p is the period between time k and $k + 1$ and p is the number of cores.

The problem is the estimation of the parameter D in the case of nonstationary task arrival processes. This means that statistical properties of the task arrival process are changing over time. The task arrival has been modeled this way since there is no guarantee that user requirements will not change over time.

To estimate statistical parameters (mean value) of this process, we use a *short-time discrete-time Fourier transform* (DSTFT) using a Kaiser window (see [28]). The DSTFT is a Fourier-related transform used to determine the sinusoidal frequency and phase content of local sections of a signal as it changes over time. It is a Fourier transform of the original signal multiplied by a windowing function. Among all the possible windowing functions, we decided to use a Kaiser window, because it is a relatively simple approximation of the prolate spheroidal function that satisfies the maximum energy concentration theorem [30].

Thus, an estimate of the parameter D is given by the following equation:

$$D = \frac{\sum_{i=0}^{N} w_\beta(i) \cdot X_i}{\sum_{i=0}^{N} w_\beta(i)} \tag{5.17}$$

where N is the number of past time slots considered and $w_\beta(i)$ is a Kaiser window of parameter β and length N. Figure 5.13 shows the shape of this type of windowing function for different values of N and β. As can be noted, for $\beta = 0$, D is the mean of the last N X_i.

The parameters N and β are left to the designer. The optimum size of these parameters depends on the way the task arrival process changes its mean and standard

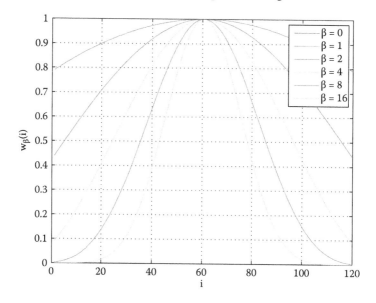

FIGURE 5.13   Kaiser window function for $N = 120$ and different values of β.

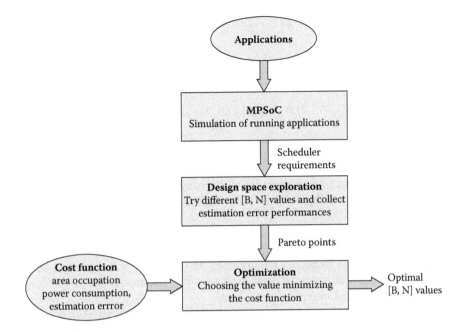

FIGURE 5.14 Block diagram description of the method used to derive β and *N* parameters from real applications or benchmarks.

deviation over time. Moreover, it depends also on the memory that is allocated to keep track of the history of the task arrival process. For slowly varying processes to have a large window size, *N* is good since this way the prediction error is small. For fast varying processes, to have a small *N* is better.

There are two fundamental parameters that are left to the designer: window width (history) *N* and the parameter β that determines the shape of the Kaiser window. To derive these parameters from real applications or benchmarks, we propose the method described in the block diagram of Figure 5.14.

Given some applications, they are run on the MPSoC. Scheduler requirements for every time window are collected and stored. After that, a design space exploration is made and the performance of different β and *N* parameters is tested.

After that, an optimization is made based on both Pareto points obtained at the end of the design space exploration and a predefined cost function. This cost function takes into account how power consumption, area occupation, and the estimation error are important in the optimization process. Finally, the result of this is the pair of β and *N* that will best fit specified needs.

5.2.3.4 Polynomial Least Squares Workload Prediction

This technique makes the prediction by using a linear model to perform a best *d*th-order polynomial fit, because the prediction length *L* for this application is short and ranges usually from one to nine samples.

$$\check{A} = \begin{pmatrix} 1 & 0 & 0 & 0 & ... & 0 \\ 1 & 1 & 1 & 1 & ... & 1 \\ 1 & 2 & 4 & 9 & ... & 2^d \\ . & . & . & . & ... & . \\ . & . & . & . & ... & . \\ . & . & . & . & ... & . \\ 1 & N+L & (N+L)^2 & (N+L)^3 & ... & (N+L)^d \end{pmatrix}$$

FIGURE 5.15   Structure of matrix $\check{A}$.

The polynomial fit is performed by minimizing the error within the observed window of temperatures, by using the following function:

$$\left\| \mathbf{w} - \check{A}\mathbf{x} \right\|_2^2 \tag{5.18}$$

where $\mathbf{w}_t$ contains the frequency requirements $\forall t = 1, ..., N$, N is the length of the observation window of historical data. Vector $\mathbf{x}_t \; \varepsilon \; \mathfrak{R}^{d+1}$ and matrix $\check{A} \; \varepsilon \; \mathfrak{R}^{d+1 \times (N+L)}$ are used in the polynomial interpolation process.

Equation 5.18 can be solved as a least squares minimization problem to derive vector $\mathbf{x}$. The prediction on the future workload requirement is performed by assuming that the linear model just derived will hold for the next L data samples.

Assuming this assumption holds, the future workload requirement is given by the following equation:

$$\mathbf{w}_t = \check{A}\mathbf{x}_t, \; \forall \; N \le t \le N + L \tag{5.19}$$

where $\mathbf{w}_t \; \varepsilon \; \mathfrak{R}^p$ for $t > N$ is the predicted workload requirement at time t. We tested the predictor on the benchmarks and achieved good accuracies for short-term forecasts (L ranging from 1 to 9).

The structure of matrix $\check{A}$ is shown in Figure 5.15. As can be noted, the first row of matrix $\check{A}$ considers only the first component of vector $\mathbf{x}$ since the first entry of the row is the only nonzero element. The second row considers all the history contained in vector $\mathbf{x}$, all with the same weight. The third row is a weight function that considers elements of vector $\mathbf{x}$ according to a quadratic weight. All rows of matrix $\check{A}$ follow the same trend, including the last row that weights entries of vector $\mathbf{x}$ with a dth polynomial function.

Values of N (the length of the observation window) and d (the order of the polynomial fit) that provide the best prediction depend on the workload requirement (task arrival process) statistical properties. These kind of processes are usually nonstationary and depend on the interaction between the user and the MPSoC itself. For the aforementioned reasons, to choose these parameters, we suggest using empirical studies.

5.2.4   HEAT PROPAGATION

We divided the chip floor plan into grid cells with cubic shapes. Each functional unit in the floor plan can be represented by one or more thermal cells in the silicon layer. The thermal behavior of the MPSoC is computed through the interaction of multiple discrete-time differential equations modeling the thermal interactions between neighboring cells. A graphical representation of an MPSoC structure is presented in Figure 5.16. In Figure 5.16, the gray and black blocks represent the silicon and copper layers, respectively. The ambient temperature is modeled as a layer with uniform temperature and infinite thermal capacity. The marking inside the silicon cell represents the cell's power dissipation. At any moment in time, the temperature change of each block due to its neighbors is given by the temperature difference between the two blocks, multiplied by the constant that labels each pair of arrows.

The thermal model is formed by considering the heat conductances **G** and capacitances **C** of the cells [29, 40]. The differential equation modeling the heat flow is given by

$$\mathbf{C} \cdot \frac{\partial t(\theta)}{\partial \theta}\bigg|_{\theta=\tau} = -\mathbf{G}(\tau) \cdot \mathbf{t}_\tau + \mathbf{p}_\tau \qquad (5.20)$$

where $\mathbf{t}_\tau$ is the temperature vector at time $\theta = \tau$. In this model we assume that lateral heat capacitances are negligible. For this reason, matrix **C** is diagonal with the entries representing the thermal capacitance of the cells (in Joules/Kelvin). Matrix **G** is the thermal conductance matrix (the conductance values have units of Watts/Kelvin). The silicon thermal conductivity is shown in Figure 5.17. As analyzed by Ramalingam et al. [33], it varies with temperature in an approximately linear fashion. For this reason, matrix **G** is a function of the thermal profile of the MPSoC at time τ.

We want to represent the thermal model using a standard state-space representation [13]. This representation uses a linear, time-invariant discrete-time system model. Since the original thermal model is nonlinear, and coefficients are temperature dependent [29], we need to linearize the solution of the differential equation (5.20) modeling the heat flow inside the MPSoC system. The rationale behind

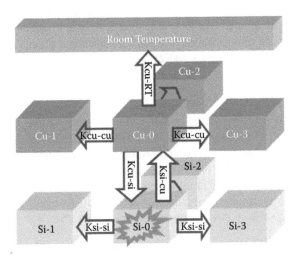

FIGURE 5.16   Modeling of the heat transfer inside a 2D-MPSoC.

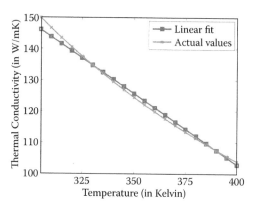

FIGURE 5.17   Silicon thermal conductivity and linear fit.

it is described in Zanini et al. [46], Skadron et al. [40], and Paci et al. [29]. In the following subsections we will describe mathematically how the thermal model of the MPSoC can be represented by the following equation:

$$\mathbf{t}_{\tau+1} = \mathbf{A}\mathbf{t}_{\tau} + \mathbf{B}\mathbf{p}_{\tau} + \mathbf{w} \qquad (5.21)$$

This equation expresses the temperature profile vector at time step $\tau + 1$, $\mathbf{t}_{\tau+1}$ ε $\mathfrak{R}^n$, as a function of the current thermal profile at step τ and the input vector $\mathbf{p}_{\tau}\varepsilon$ $\mathfrak{R}^{p+z}$. This model is discrete-time with time step equal to $\Delta\tau$. The total number of cells in all layers of the 3D-MPSoC structure is n, the total number of independent processing units is p, and the total number of independent cooling liquid flow rates is z. Matrices $\mathbf{A}$ ε $\mathfrak{R}^{n \times n}$ and $\mathbf{B}$ ε $\mathfrak{R}^{n \times (p+z)}$ describe the heat propagation properties of the MPSoC, while $\mathbf{w}$ ε $\mathfrak{R}^n$ takes into account the fact that the MPSoC is at room temperature. If we refer to all temperatures as offset from the room temperature, we can omit this vector. At time τ, the temperature of the next simulation step of cell i, i.e., $(\mathbf{t}_{\tau+1})_i$, can be computed thanks to Equation 5.21. The first p entries of vector $\mathbf{p}$ are the power consumptions for each of the p independent processing units, while the remaining z entries are the normalized cooling flow rates for each of the z independent microchannels.

5.2.4.1   First-Order ODE Solvers

A first-order ordinary differential equation (ODE) solver is the simplest tool for solving the system of differential equations modeling the MPSoC. Because of its simplicity, it has been used by many state-of-the-art thermal simulators like the earliest versions of HotSpot [40] or real-time thermal emulation frameworks targeting embedded SoCs [29]. This integration method has been used in early thermal policies embedding a thermal profile model of the MPSoC (see [27] for more details). The *forward Euler* (FE) method is described by Equation 5.22:

$$\mathbf{t}_{\tau+\Delta\tau} = \mathbf{t}_{\tau} + \Delta\tau \cdot \left.\frac{\partial \mathbf{t}(\theta)}{\partial\theta}\right|_{\theta=\tau} \qquad (5.22)$$

where $\mathbf{t}_\tau$ and $\mathbf{t}_{\tau+\Delta\tau}$ are the vectors containing the temperature of any thermal cell composing the MPSoC, respectively, at time τ and $\tau + \Delta\tau$. θ is the time, $\Delta\tau$ is the simulation step size, and $\left.\dfrac{\partial \mathbf{t}(\theta)}{\theta}\right|_{\theta=\tau}$ is the vector containing the temperature rate of change at time τ.

The temperature rate of change at time τ in Equation 5.20 is given by

$$\left.\frac{\partial \mathbf{t}(\theta)}{\partial\theta}\right|_{\theta=\tau} = -\mathbf{C}^{-1}\mathbf{G}(\tau)\cdot\mathbf{t}_\tau + \mathbf{C}^{-1}\mathbf{p}_\tau \tag{5.23}$$

while, according to Equation 5.22, we have that

$$\left.\frac{\partial \mathbf{t}(\theta)}{\partial\theta}\right|_{\theta=\tau} = \mathbf{A}'\sim(\tau)\cdot\mathbf{t}_\tau + \mathbf{B}'\cdot\mathbf{p}_\tau + \mathbf{w}' \tag{5.24}$$

By solving the system composed by Equations 5.23 and 5.24, we have that

$$\mathbf{A}'(\tau) = -\mathbf{C}^{-1}\mathbf{G}(\tau) \tag{5.25}$$

$$\mathbf{B}' = \mathbf{C}^{-1} \tag{5.26}$$

where matrix $\mathbf{A}'(\tau)$ expresses the part of the on-chip temperature spreading process that depends only on the cell's temperature. This matrix models the thermal conductances and capacitances network of Figure 5.16 in the matrix form, except K_{cu-RT}, which needs to be modeled separately. Matrix $\mathbf{B}'$ is a matrix where $\mathbf{B}'_{i,j}$ contains the conversion factor between the power assigned to functional unit j and the temperature increase in cell i. Matrices $\mathbf{A}'(\tau)$ and $\mathbf{B}'$ contain the system dynamics that depend entirely on the current state and on the given power assignment vector $\mathbf{p}_\tau$. The part of the dynamic system that is not controllable by the input vector, such as the heat dissipation of the copper layer due to room temperature, is expressed by vector $\mathbf{w}'$. Then, by substituting Equation 5.24 into Equation 5.22, Equation 5.27 is obtained:

$$\mathbf{t}_{\tau+\Delta\tau} = \mathbf{A}(\tau)\cdot\mathbf{t}_\tau + \mathbf{B}\cdot\mathbf{p}_\tau + \mathbf{w} \tag{5.27}$$

where

$$\mathbf{A}(\tau) = \mathbf{I} + \Delta\tau\mathbf{A}'(\tau) \tag{5.28}$$

$$\mathbf{B} = \Delta\tau\mathbf{B}' \tag{5.29}$$

$$\mathbf{w} = \Delta\tau\mathbf{w}' \tag{5.30}$$

Equation 5.27 is a time-varying state-space representation modeling the thermal behavior of the MPSoC using a first-order ODE solver. The computation here is simple and requires only a matrix multiplication. The total number of cells in all layers of the 3D-MPSoC structure is n, the total number of independent processing units is p, and the total number of independent cooling liquid flow rates is z.

An important property of a solver is its stability. An integration method is called numerically stable if an error does not exponentially grow during the calculation of the final solution. The FE method has potential stability problems when the chosen time step for the thermal simulations is large, as shown in the literature [9], which will be explored and addressed in this chapter.

A first-order ODE method that is unconditionally stable is the *backward Euler* (BE) method. This method is described by Equation 5.31:

$$\mathbf{t}_{\tau+\Delta\tau} = \mathbf{t}_\tau + \Delta\tau \cdot \left. \frac{\partial \mathbf{t}(\theta)}{\partial \theta} \right|_{\theta=\tau+\Delta\tau} \tag{5.31}$$

Assuming $\mathbf{A}'(\tau) \simeq \mathbf{A}'(\tau+\Delta\tau)$, and using Equation 5.32,

$$\left. \frac{\partial \mathbf{t}(\theta)}{\partial \theta} \right|_{\theta=\tau+\Delta\tau} = \mathbf{A}'(\tau+\Delta\tau) \cdot \mathbf{t}_\tau + \Delta\tau + \mathbf{B}' \cdot \mathbf{p}_\tau + \mathbf{w}' \tag{5.32}$$

we can obtain Equation 5.33 in a discrete-time domain:

$$\mathbf{t}_{\tau+\Delta\tau} = \mathbf{A}(\tau) \cdot \mathbf{t}_\tau + \mathbf{B}(\tau) \cdot \mathbf{p}_\tau + \mathbf{w}(\tau) \tag{5.33}$$

where

$$\mathbf{A}(\tau) = [\mathbf{I} - \Delta\tau\mathbf{A}'(\tau)]^{-1} \tag{5.34}$$

$$\mathbf{B}(\tau) = [\mathbf{I} - \Delta\tau\mathbf{A}'(\tau)]^{-1} \Delta\tau\mathbf{B}' \tag{5.35}$$

$$\mathbf{w}(\tau) = [\mathbf{I} - \Delta\tau\mathbf{A}'(\tau)]^{-1} \Delta\tau\mathbf{w}' \tag{5.36}$$

This method achieves unconditional stability at the cost of a significant increase in computational complexity with respect to the FE algorithm. The most expensive computational step for the BE method is the inverse matrix computation. It has been shown in the literature [29, 40] that the accuracy of both first-order methods is $O(\Delta\tau^2)$ [9].

5.2.4.2 Second-Order ODE Solvers

As a representative example of second-order solvers, we analyze the Crank-Nicholson (CN) method, also called trapezoidal. This integration algorithm combines FE with BE to obtain a second-order method due to cancellation of the error terms. The CN method reaches an accuracy of $O(\Delta\tau^2)$. This method can be described using Equation 5.37:

$$\mathbf{t}_{\tau+\Delta\tau} = \mathbf{t}_\tau + \frac{\Delta\tau}{2} \cdot \left[\left. \frac{\partial \mathbf{t}(\theta)}{\partial \theta} \right|_{\theta=\tau} + \left. \frac{\partial \mathbf{t}(\theta)}{\partial \theta} \right|_{\theta=\tau+\Delta\tau} \right] \tag{5.37}$$

Assuming $\mathbf{A}'(\tau) \simeq \mathbf{A}'(\tau + \Delta\tau)$, and using Equations 5.24 and 5.32, we obtain

$$\mathbf{t}_{\tau+\Delta\tau} = \mathbf{A}(\tau) \cdot \mathbf{t}_\tau \, \mathbf{B}(\tau) \cdot \mathbf{p}_\tau + \mathbf{w}(\tau) \tag{5.38}$$

where

$$\mathbf{A}(\tau) = [\mathbf{I} - 0.5 \cdot \Delta\tau \mathbf{A}'(\tau)]^{-1} [\mathbf{I} + 0.5 \cdot \Delta\tau \mathbf{A}'(\tau)] \tag{5.39}$$

$$\mathbf{B}(\tau) = [\mathbf{I} - 0.5 \cdot \Delta\tau \mathbf{A}'(\tau)]^{-1} \Delta\tau \mathbf{B}' \tag{5.40}$$

$$\mathbf{w}(\tau) = [\mathbf{I} - 0.5 \cdot \Delta\tau \mathbf{A}'(\tau)]^{-1} \Delta\tau \mathbf{w}' \tag{5.41}$$

The CN method is stable and has a higher accuracy in comparison to first-order solvers when larger simulation time steps are used. The accuracy of such second-order methods is $O(\Delta\tau^3)$ [9].

5.2.4.3 Multistep Fourth-Order ODE Solver

Numerical ODE solution methods start from an initial point and take a small step in time to find the next solution point. This process continues with subsequent steps to compute the solution. Single-step methods (such as Euler's method) refer to only one previous point and its derivative to determine the current value. Multistep methods take several intermediate points within every simulation step to obtain a higher-order method. This way, they increase the accuracy of the approximation of the derivatives by using a linear combination of these internal additional points. A multistep solver has been embedded in version 4.0 of HotSpot [40]. One particular subgroup of this family of multistep solvers is the Runge-Kutta (RK4) method, which includes a fourth-order solver. The algorithm that we use for implementing the RK4 solver employs an FE method to compute derivatives at the internal points. By using the model represented in Figure 5.16, this method is described by the following equations:

$$\mathbf{k}_1 = \Delta\tau \cdot [\mathbf{A}'(\tau)\mathbf{t}_\tau + \mathbf{B}' \cdot \mathbf{p}\tau + \mathbf{w}'] \tag{5.42}$$

$$\mathbf{k}_2 = \Delta\tau \cdot [\mathbf{A}'(\tau + 0.5\Delta\tau)(\mathbf{t}_\tau + 0.5 \cdot \mathbf{k}_1) + \mathbf{B}' \cdot \mathbf{p}\tau + \mathbf{w}'] \tag{5.43}$$

$$\mathbf{k}_3 = \Delta\tau \cdot [\mathbf{A}'(\tau + 0.5\Delta\tau)(\mathbf{t}_\tau + 0.5 \cdot \mathbf{k}_2) + \mathbf{B}' \cdot \mathbf{p}\tau + \mathbf{w}'] \tag{5.44}$$

$$\mathbf{k}_4 = \Delta\tau \cdot [\mathbf{A}'(\tau + \Delta\tau)(\mathbf{t}_\tau + \mathbf{k}_3) + \mathbf{B}' \cdot \mathbf{p}\tau + \mathbf{w}'] \tag{5.45}$$

$$\mathbf{t}_{\tau+\Delta\tau} = \mathbf{t}_\tau + \frac{1}{6} \cdot (\mathbf{k}_1 + 2\mathbf{k}_2 + 2\mathbf{k}_3 + \mathbf{k}_4) \tag{5.46}$$

Assuming $\mathbf{A}'(\tau) \simeq \mathbf{A}'(\tau + 0.5\Delta\tau) \simeq \mathbf{A}'(\tau + \Delta\tau)$, we obtain

$$\mathbf{t}_{\tau+\Delta\tau} = \mathbf{A}(\tau) \cdot \mathbf{t}_\tau + \mathbf{B}(\tau) \cdot \mathbf{p}_\tau + \mathbf{w}(\tau) \tag{5.47}$$

where

$$F = \Delta_t A' (\tau) \tag{5.48}$$

$$A(\tau) = I + \frac{1}{6}\left[6F + 3F^2 + F^3 + 0.25F^4\right] \tag{5.49}$$

$$B(\tau) = \Delta\tau \cdot \left[I + \frac{1}{6}(3F + F^2 + 0.25F^3)\right] \cdot B' \tag{5.50}$$

$$w(\tau) = \Delta\tau \cdot \left[I + \frac{1}{6}(3F + F^2 + 0.25F^3)\right] \cdot w' \tag{5.51}$$

Note that this method does not require the inverse matrix computation. In addition, like FE, this method is not unconditionally stable, since the RK4 method uses the FE for computing the rate of change of the temperature function in the internal point, and hence inherits its stability properties. The accuracy of this multistep fourth-order method can reach $O(\Delta_t^5)$ [9].

5.2.4.4 Changing the Sampling Rate

Equation 5.27 models the RC network representing the MPSoC behavior. To allow high accuracy in the discrete integration process, the sampling period between t_τ and $t_{\tau+\Delta\tau}$ has to be small (i.e., 10–200 µs). An MPSoC heating process makes relevant changes in a time range that is several orders of magnitude the simulation time step cited before. This requires many iterations of Equation 5.27, and this is costly if, for each step, there is a thermal policy computation associated with it. To increase the sampling rate to $\Delta'\tau = v \cdot \Delta\tau$ without changing the value of $\Delta\tau$, the following mathematical derivation is presented.

Assuming that the input p does not change during $\Delta'\tau$ as well as matrices A and B and vector w, we have that Equation 5.27 turns into Equation 5.52:

$$t_{\tau+\Delta\tau} = A \cdot t_\tau + B \cdot p + w \tag{5.52}$$

By iterating Equation 5.52, we have that

$$t_{\tau+2 \cdot \Delta\tau} = A \cdot [A \cdot t_\tau + B \cdot p + w] \cdot t_\tau + B \cdot p_\tau + w \tag{5.53}$$

By iterating Equation 5.53 using Equation 5.52, we have that at step 3:

$$t_{\tau+3 \cdot \Delta\tau} = A \cdot [A \cdot [A \cdot t_\tau + B \cdot p + w] \cdot t_\tau + B \cdot p + w] t_\tau + B \cdot p_\tau + w \tag{5.54}$$

The way Equation 5.52 has evolved to Equation 5.54, we derive the following generalization that holds for any step v:

$$t_{\tau+v \cdot \Delta\tau} = A^v t_\tau + \sum_{i=0}^{v-1} A^i Bp + \sum_{i=0}^{v-1} A^i w \tag{5.55}$$

Exploiting matrix series properties we have that

$$t_{\tau+v\,\Delta\tau} = A^{\,v}t_\tau + [(I-A)^{-1}(I-A^v)]Bp + [(I-A)^{-1}(I-A^v)]\,w \qquad (5.56)$$

where I is the identity matrix with the same size of the square matrix A. By comparing Equation 5.56 with Equation 5.52, we have that

$$t_{\tau+v\,\Delta\tau} = A_v t_\tau + B_v p + w_v \qquad (5.57)$$

where

$$A_v = A^v$$
$$B_v = [(I-A)^{-1}(I-A^v)]B \qquad (5.58)$$
$$w_v = [(I-A)^{-1}(I-A^v)]w$$

It is important to notice that this derivation holds only if the input p does not change during $\Delta'\tau = v \cdot \Delta\tau$ as well as matrices A and B and vector w. The assumption that matrix A does not change is an approximation that is close to reality, as the thermal profile does not make big variations in $\Delta'\tau$.

5.2.5   LIQUID COOLING

In 3D stacks, cooling cannot be handled and managed by conventional air cooling methods over the stack surface because of the large heat propagation. Interlayer liquid cooling is a potential solution to address thermal problems. Brunschwiler et al. [6] reported that liquid cooling solutions offer a higher heat removal capability in comparison to air and to the possibility to extract heat at various layers of the stack.

5.2.5.1   Straight and Bent Microchannels

There are several ways to support liquid cooling, e.g., by adding/inserting to the stack a plate with built-in microchannels or by etching a porous media structure between the tiers of the 3D stack [6, 7]. Experiments have shown that when a coolant fluid is pumped through the microchannels, up to 3.9 KW/cm³ [7] of heat can be extracted.

Porous media structures can be designed with different forms according to the TSV spacing requirements and the desired fluidic path [41, 42]. Figure 5.18 shows a planar view of two different structures. Although these structures use microchannels to guide the fluid, one of them uses straight channels with two ports (Figure 5.18a), while the other exploits bent channels and four ports (Figure 5.18b). In the following we will refer to these structures as straight and bent channels. Using multiport bent channels is more beneficial than using straight channels, if the straight channel length is longer than the thermal developing length of the fluid [7]. Moreover, bent channels have different lengths, thus enabling different liquid flow rates between different channels.

Overall, thermal management of a 3D stack is achieved by a combination of active control of on-chip switching rates (the heat source) and active interlayer cooling with

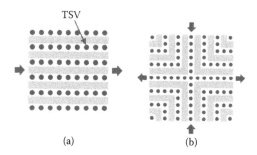

TSV

(a) (b)

FIGURE 5.18 Top view of (a) two-port and (b) four-port microchannel fluid delivery architecture compatible with area-array interconnects.

pressurized fluids (the heat sink). It is important to remember that the cooling system requires one or more pumps to circulate the fluid, as well as a heat exchanger to cool the fluid. The latter may be passive (e.g., fin structure) or active (e.g., fan). At any rate, a relevant part of the system energy spent for cooling is due to the pump [36] and a minor part by the exchanger.

5.2.5.2 Flow Rate and Channel Length

The relation between the flow rate Fl and the channel length L is as follows:

$$Fl = w_{ch} \cdot h_{ch} \cdot v_{bulk} \tag{5.59}$$

$$v_{bulk} = \frac{v_{darcy}}{\varepsilon} \tag{5.60}$$

$$v_{darcy} = \frac{\kappa}{\mu} \cdot \nabla P \tag{5.61}$$

$$\nabla P = \frac{\Delta P}{L} \tag{5.62}$$

where v_{bulk} is the actual fluid velocity and v_{darcy} is the fluid velocity multiplied by the cavity porosity. Parameters used in Equations 5.59–5.62 are shown in Table 5.1. Hence, the flow rate and channel length are inversely proportional; i.e., the shorter the channel length is, the higher the flow rate is. We validate the flow velocity obtained for each channel by comparing the analytical model in Equations 5.59–5.62 with the experimental values shown in [7]. As Figure 5.19 shows, the proposed analytical model provides us an acceptable method to calculate the flow rate for different channel lengths.

5.2.5.3 Thermal Capability and Pumping Power

Since we use varying flow rate as a control variable for energy-efficient thermal management, it is crucial to study the thermal capability of interlayer liquid cooling with respect to different pumping power values. Thus, each pumping power value is

TABLE 5.1

Parameter Definitions Used to Relate the Flow Rate to the Channel Length

Parameter	Definition
w_{ch}	Channel width (50 μm)
h_{ch}	Channel height (100 μm)
ε	Cavity porosity (0.5)
κ	Cavity permeability ($7.17E-11$ m²)
μ	Dynamic viscosity ($1E-3$ Pascal · s)
ΔP	Pressure difference between the inlet and outlet ports (1 bar)

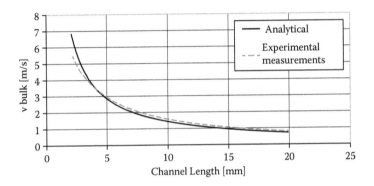

FIGURE 5.19 Comparison of the fluid flow velocity in different channel lengths between the analytical method (Equations 5.59–5.62) and the experimental results shown in Brunschwiler et al., Validation of the porous-medium approach to model interlayer-cooled 3D chip stacks. In *3DIC*, 2009.

translated to a specific flow rate in our system. First, we use Bernoulli's equation to describe the pump power P_{pump} as follows:

$$P_{pump} = \frac{\Delta P \cdot Fl}{\zeta} \qquad (5.63)$$

where ΔP is the pressure difference required, Fl is the fluid flow rate, and ζ is the pumping power efficiency. We use $\zeta = 0.7$, as it is a normal pump efficiency value [11, 20]. Since there is a linear relation between the pressure difference and the flow rate injected in the stack (Equations 5.59–5.62), we can say that $P_{pump} \propto Fl^2$.

Next, we define the thermal capability of interlayer liquid cooling as the maximum heat flux absorbed by the fluid to keep the maximum temperature within the stack below 85°C. To estimate this thermal capability, we use 3D-ICE [41] to record the maximum temperature of the stack at different thermal dissipation values, and with different flow rates. We limit the maximum flow rate injected to be the one at $\Delta P = 1$ bar, since it is the maximum safe pressure requirement within the stack [6].

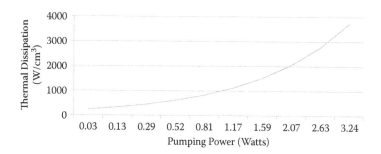

FIGURE 5.20 Rate of change of thermal capability of interlayer liquid cooling *TC* with respect to pumping power P_{pump}.

Therefore, Figure 5.20 shows the amount of minimum pumping power applied to keep the maximum temperature of the stack below 85°C, at different thermal dissipation rates.

5.2.5.4 Flow Rate and Heat Extraction

The amount of heat r_i extracted in cell i by the fluid in the microchannel controlled by pump j can be approximated by the following equation:

$$r_i = m_j \cdot \gamma_{i,j} \cdot (t_i - t_{fluid}) \tag{5.64}$$

where the fluid temperature is t_{fluid}, t_i is the temperature of cell i, and $\gamma_{i,j}$ is the constant modeling of the channel heat extraction properties. Vector $\mathbf{m} \; \varepsilon \; \Re^z$ is the normalized amount of heat that can be extracted for each of the z independent pumps.

Thus, by varying vector $\mathbf{m}$, the cooling power (flow rate of the cooling liquid) is varied to achieve the desired heat extraction. In our model, we used the temperature mapping from [41] to derive $\gamma_{i,j}$.

Experiments have shown that by updating $\gamma_{i,j}$ every time the policy is applied (10 ms in our simulation setup), our approximation leads to a maximum error up to ± 5%.

5.3 SENSOR PLACEMENT

5.3.1 INTRODUCTION

In the thermal model described in previous sections, a state is required for every block of the floor plan, because we need n states to store n temperature values. This requirement is expensive in terms of computational cost. The higher the number of states modeling the MPSoC, the higher is the number of sensors required for its state estimation This could be a problem in case of a detailed model of a complex 3D-MPSoC including liquid cooling.

This approach finds the best locations inside the 3D-MPSoC where thermal sensors can be placed using a greedy technique. The advantage is an efficient method to solve both the sensor placement and the model order reduction of the MPSoC.

Complex MPSoC models are indeed reduced in complexity, with an inaccuracy in the order of few percent. As a result, the thermal model is simpler that the original one, and the computational cost of the thermal management algorithm will be reduced.

The block diagram of the algorithm is presented in Figure 5.21. The proposed methodology consists of two phases: a design-time phase and a runtime phase.

During the design phase the thermal management system model is defined. The reduced-order MPSoC thermal model and the sensor placement are the outputs of this off-line phase. The concept behind the proposed sensor placement technique is based on an analysis of the balanced state-space realization of the 3D-MPSoC system and its Hankel singular values' decay rate (see [13, 21] for more details). The number of states of the reduced-order model is fixed according to user designer accuracy requirements, and a specific location is assigned to each sensor.

During the runtime phase the reduced-order system state vector **x** is estimated thanks to a simple state estimator (i.e., Kalman filter) and measurements coming from thermal sensors. Then, this information is used by the thermal model to perform the optimization on the reduced-order 3D-MPSoC model predefined in the design-time phase.

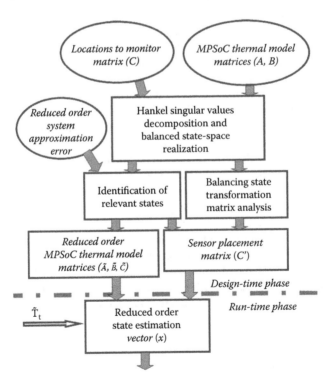

FIGURE 5.21 Block diagram of the method proposed in Zanini, Atienza, and De Micheli, A combined sensor placement and convex optimization approach for thermal management in 3D-MPSoC with liquid cooling, *Integration, the VLSI Journal*, 46(1): 33–43, 2013.

5.3.2   From Structure-Centric to Energy-Centric

First, an accurate 3D-MPSoC thermal model is created according to the model presented in Section 5.3.1. This will determine matrices $\mathbf{A}$ and $\mathbf{B}$ according to Equation 5.65. We assume here that all temperatures are offsets from room temperature. This way we can omit vector $\mathbf{w}$ from Equation 5.21.

$$\mathbf{t}_{\tau+1} = \mathbf{A}\mathbf{t}_\tau + \mathbf{B}\mathbf{p}_\tau \tag{5.65}$$

Locations that the policy needs to monitor to ensure safe working conditions are determined by the following relation:

$$\tilde{\mathbf{t}}_\tau = \mathbf{C}\mathbf{t}_\tau \tag{5.66}$$

Equation 5.66 describes the choice of relevant locations to monitor inside the MPSoC. Matrix $\mathbf{C} \, \varepsilon \, \mathbf{B} = \{0,1\}^{s \times n}$ is a selection matrix. In this model we assume that we want to control locations on the silicon layer of each tier. We do this to ensure a full MPSoC temperature control in every location containing an active device on the silicon layer. We assume that s is the total number of those locations. Namely, $c_{i,j}$ is equal to 1 if thermal sensor i is located inside cell j.

To determine the states with negligible contribution to the MPSoC thermal dynamics, the system is balanced using a Gramian-based balancing of state-space realizations [21]. This technique computes a balanced state-space realization for the stable portion of the system. For stable systems, the output is an equivalent system for which the controllability and observability Gramians are equal and diagonal, their diagonal entries forming the vector $\mathbf{g} \, \varepsilon \, \mathfrak{R}^n$ of Hankel singular values. These values provide a measure of energy for each state in the system. If the corresponding Hankel singular value for a certain state is a relatively small number, this means that that state has a small influence in the dynamic of the system. The second output of the Gramian-based balancing [21] is the balancing state transformation matrix $\mathbf{T} \, \varepsilon \, \mathfrak{R}^{n \times n}$ that converts the original system into the balanced one.

The rationale behind this operation is to change the 3D-MPSoC thermal model system perspective. The original model belongs to a geometric and physical view of the 3D-MPSoC where states are related to physical properties. The new model generated by the Gramian-based transformation is energy-centric, and every state is a heat propagation dynamic. This representation emphasizes how much the thermal dynamic represented by that specific state is relevant to the heat propagation response of the system. The ith row of the conversion matrix $\mathbf{T}$ describes the contribution that the temperature of each thermal cell in the original model gives to the ith most important (in terms of energy) thermal dynamic of the new generated system.

5.3.3   Identification of Relevant States

Here we elaborate information related to Hankel singular values vector $\mathbf{g}$. Figure 5.22 shows the state energy distribution for the case study described in [48]. As Figure 5.22

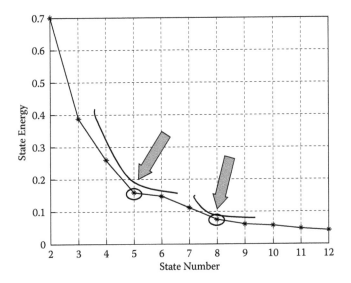

FIGURE 5.22 Decay rate analysis for the normalized energy related to Hankel singular values for the case study described in Zanini, Atienza, and De Micheli, A combined sensor placement and convex optimization approach for thermal management in 3D-MPSoC with liquid cooling, *Integration, the VLSI Journal*, 46(1): 33–43, 2013. Arrows point to change in the decay rate.

shows, the energy magnitude drops quite fast, and most of the states give almost negligible contributions to the input-output response of the system. To define a threshold level to distinguish between relevant and not relevant states, we look at the rate of decay of the state energy.

Figure 5.22 shows the decay rate for the normalized energy related to Hankel singular values for the case study described in [48]. Arrows point to change in the decay rate. To identify transition points, we look at peaks in the third derivative of the function defined by vector **g**. In Figure 5.22 they are highlighted with circles. All these points represent a set of possible threshold points to distinguish between relevant states and negligible states. This means that by adding points after these transition points, the advantage of adding states would be smaller in terms of reducing the approximation error.

Figure 5.23 shows the model approximation error in percentage versus number of states in the reduced model. As Figure 5.23 shows, the decay rate is fast. It is important to notice that only threshold points have been considered in this plot. They are marked with *. Results show that an approximation error of $6 \cdot 10^{-2}$ in percentage can be achieved with only 20 states, and 180 can be easily discarded with a reduction factor of 10× for the given accuracy. It is also important to notice that it does not make much sense to go for higher accuracies because inaccuracies in the silicon, in the power model, or in thermal sensors will add uncertainty in the results.

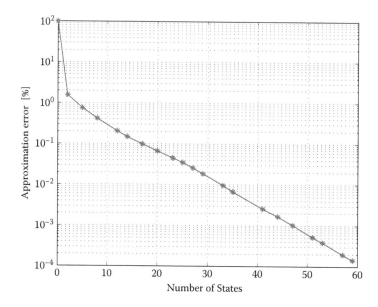

FIGURE 5.23   Model approximation error (%) versus number of states in the reduced model for the case study described in Zanini, Atienza, and De Micheli, A combined sensor placement and convex optimization approach for thermal management in 3D-MPSoC with liquid cooling, *Integration, the VLSI Journal*, 46(1): 33–43, 2013.

5.3.4   BALANCED STATE TRANSFORMATION ANALYSIS

Here we elaborate information related to conversion matrix **T**. The ith row of **T** describes the contribution that the temperature of each thermal cell in the original model gives to the ith most important (in terms of energy) thermal dynamics of the new generated system. For this reason, at this stage for each row i of **T**, we identify the most relevant component in absolute value. We call this component j. This means that if we place a sensor in the jth cell in the original model, among all the possible sensor locations, this position would be the one that will contribute more in terms of energy to the ith most important thermal dynamic of the new generated system.

Figure 5.24 shows the most relevant component identifying each state of the new thermal model. Horizontal lines delimit one layer from another. They are placed at multiples of 25 because each layer according to the case study we are considering consists of 25 cells (see [48] for more details). As Figure 5.24 shows, there are no sensors in the copper layer. The reason is because we decided to add this technological limitation as a constraint in the definition of the most relevant sensor location for each state.

5.3.5   REDUCED-ORDER MODEL AND SENSOR PLACEMENT

At this stage the user-defined parameter that is missing to complete the sensor placement is the desired accuracy of the reduced-order model. If we accept an approximation error of $6 \cdot 10^{-2}$ in percentage, we fix the number of states to 20.

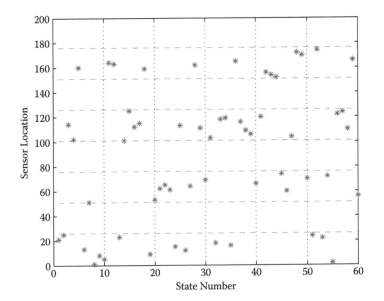

FIGURE 5.24 Sensor location according to the most relevant component identifying each state of the new thermal model. Horizontal lines delimit one layer from another.

By doing this operation, we reduce by a factor of 10 the number of states in the model, and so the computational complexity of Equation 5.65.

At this point a new reduced-order model is obtained from the original one after the balancing using a Gramian-based balancing of state-space realizations [21]. States corresponding to Hankel singular values smaller than a predefined threshold (in our case we selected the 20th) are discarded. Thus, the full MPSoC thermal model is now described by the following system of equations:

$$\mathbf{x}_{\tau+1} = \tilde{\mathbf{A}}\mathbf{x}_{\tau} + \tilde{\mathbf{B}}\mathbf{p}_{\tau} \tag{5.67}$$

$$\tilde{\mathbf{t}}_{\tau} = \tilde{\mathbf{C}}\mathbf{x}_{\tau} \tag{5.68}$$

where matrix $\tilde{\mathbf{A}} \in \mathfrak{R}^{l\times l}$ and matrix $\tilde{\mathbf{B}} \in \mathfrak{R}^{l\times p}$. The number of states of the new thermal model is l, and p is the number of inputs in the MPSoC model. Equation 5.67 describes the state update for the reduced-order model of the MPSoC. This equation is analogous to Equation 5.65. The only difference is that, in this case, the states do not represent directly temperature values inside each cell. Matrix $\tilde{\mathbf{C}} \in \mathfrak{R}^{s\times l}$ in Equation 5.68 relates the value of the states to temperature in s specific locations (every cell in all silicon layers) inside the MPSoC. This equation is analogous to Equation 5.66 and describes how the temperature measurements can be derived from the state vector $\mathbf{x}$. In our case, we were interested in knowing all the cell temperatures of every silicon layer; for this reason, $s = 100$.

The purpose of sensor placement is to get reliable information on the 3D-MPSoC thermal profile. The reason is because every time any policy is applied, it operates on reliable thermal profile temperature values. The key for this is to obtain the state vector **x**. In Section 5.3.3, the balancing state transformation matrix **T** converts the original system into the balanced one. Thanks to this matrix, to obtain the estimate of the reduced state vector **x**, it is sufficient to multiply the thermal profile by matrix **T**.

For the system identified by Equation 5.65, it means that we are able to reconstruct completely the thermal profile of the chip given the inputs only by looking at the measurements coming from the sensors, placed in locations specified by the matrix C':

$$\tilde{\mathbf{t}}_\tau = \mathbf{C}'\mathbf{t}_\tau \tag{5.69}$$

Matrix $\mathbf{C}' \; \varepsilon \; \mathbb{R}^{s' \times l}$ in Equation 5.69 is a selection matrix that describes the sensor placement inside the 3D-MPSoC. This means that we are assuming to have in the output vector s' distinct temperature measurements coming from s' distinct cells every T_s seconds, where T_s is the sensors' sampling period. The rank of the observability matrix **Q** expresses the number of states that can be reconstructed from the measurement vector $\tilde{\mathbf{t}}_\tau$. The observability matrix **Q** is expressed by the following equation (see [13]):

$$\mathbf{Q} = [\mathbf{C}'; \; \mathbf{C}' \; \mathbf{A}; ...; \; \mathbf{C}' \; \mathbf{A}^{n-1}] \tag{5.70}$$

If the rank of matrix **Q** equals n, the state vector **x** can be reconstructed completely from the measurements, the input vector **p**, and matrices **A** and **B** identifying the thermal dynamics of the system (see Equation 5.65).

The problem of selecting the right placement of thermal sensors to both minimize the number of sensors and maximize observability is the problem of choosing the matrix **C'** with the minimum number of rows that makes the rank of the observability matrix **Q** equal n. Given an MPSoC model, this problem depends on the location and the number of sensors inside a floor plan (matrix **C'**) and the sensor sampling period T_s.

To choose the sensor placement, we used the information about the locations that contribute most to each of the states in the balanced model. The algorithm is a greedy technique that adds a sensor position according to the placement suggested in Section 5.3.4. Figure 5.24 shows the most relevant component identifying each state of the new thermal model, and the algorithm starts from the most relevant state (state number 1) and goes on adding sensors until the rank of the observability matrix equals the rank of **A**. Figure 5.25 shows the percentage of states that can be estimated versus the number of sensors placed. As can be noted, for each sensor that is placed as suggested by Figure 5.24, there is an increase in the number of states of the reduced-order model that can be estimated. The complete estimation is achieved with 25 sensors for the case study described in [48]. This means that with only 25 sensors it is possible to estimate the thermal profile of the 3D-MPSoC. This means that we achieved a reduction of a factor 8 in the number of required sensors. Figure 5.26 shows the resulting placement assuming a sensor's sampling frequency T_s of 1 ms.

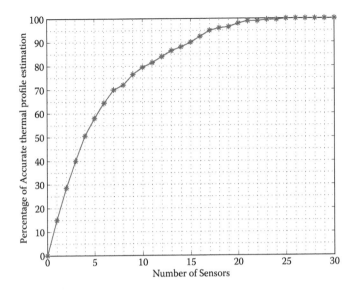

FIGURE 5.25 Sensor placement algorithm: percentage of accurate temperature estimation according to the number of sensors placed with the proposed methodology.

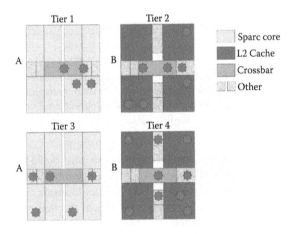

FIGURE 5.26 Sensor placement for the case study described in Zanini et al. [48] with a sensor (marked as red stars on the floor plan) sampling frequency T_s of 1 ms.

5.4 DISTRIBUTED THERMAL MANAGEMENT

5.4.1 HIERARCHICAL STRUCTURE

The structure of a hierarchical thermal management system is shown in Figure 5.27: the 3D-MPSoC architecture is partitioned into p tiers (or layers) where, without loss of generality, each tier is a subsystem of the 3D-MPSoC. In our exploration, we define a tier as a complete layer.

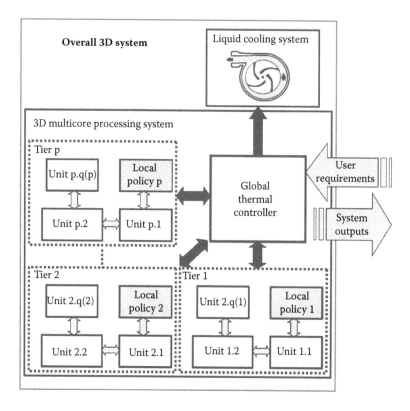

FIGURE 5.27 Structure of a hierarchical thermal management system. For more details see Zanini, Sabry, Atienza, and De Micheli, Hierarchical thermal management policy for high-performance 3D systems with liquid cooling. In *IEEE Journal on Emerging and Selected Topics in Circuits and Systems*, 1(2): 88–101, 2011.

Moreover, any tier consists of several units. These units could be cores, memory storage units, or other computational units (e.g., application-specific integrated circuit (ASIC) or custom hardware blocks). Then, the units inside each tier, say tier i, are partitioned into $q(i)$ frequency islands, and a local thermal controller manages the $q(i)$ islands, i.e., sets the frequencies and voltages to all (controllable) components inside the tier. Objectives of local controllers include preventing hot spots and minimizing undone workload. Specific requirements (e.g., workload) come from a centralized unit (i.e., the global thermal controller in Figure 5.27), which is responsible for the holistic coordination of the p local thermal controllers, and which regulates the heat extraction of the cooling system by setting the pressure of the coolant liquid (by controlling the cooling pump or the controlling valve).

This hierarchical structure is crucial for scalability and feasibility of large MPSoCs [12]. Indeed, by using this hierarchical approach, we can significantly simplify the function and overhead of the global controller by using local thermal controllers. Moreover, this structure enables the global and local controllers to be executed with different rates, e.g., the optimization of the global controller can be executed at least one order of magnitude less frequently than the local regulators. The global controller manages the pumping flow rate, which is a much slower process than DVFS.

5.4.2 RUNTIME INTERACTION: GLOBAL AND LOCAL CONTROLLERS

The communication protocol between the local controllers and the global one is shown in Figure 5.28. Initially, the global controller receives a workload requirement from the scheduler as well as a data vector containing the workload fulfillment status in each specific tier from all the p local controllers. This data vector contains two pieces of information: (1) the maximum temperature measured on-line in the corresponding tier and (2) the already executed workload. Indeed, this last information provides the global controller with an overview about how well the local controllers are performing in trying to fulfill overall requirements.

Moreover, as the workload fulfillment data from all the local controllers are collected and processed, the global unit splits the overall workload into p components. Hence, for each local controller, the global unit sets the amount of workload it has to execute. It is important to notice that the controller does not perform detailed task assignment, but just sets individual targets for each tier to satisfy the overall workload. The pressure of the coolant liquid is set during this process by the global controller, which performs this operation periodically, with a period of T_{GC}. Once these

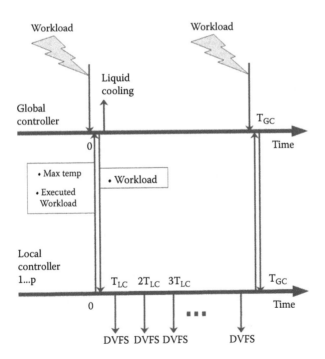

FIGURE 5.28 Communication protocol between the global and local controllers of the proposed method.

tasks are performed, the global controller stays still for the rest of the period T_{GC}. Concurrently, each local controller sets periodically the DVFS value of all related islands, but with another period T_{LC}, such that $T_{GC} = n \cdot T_{LC}$, $n \in Z^+$. The local controllers manage independently the corresponding subsystems, and they can communicate with the global thermal management unit only once in the period T_{GC}.

5.4.3 DESIGN AND IMPLEMENTATION

The design and implementation of the proposed management scheme consists of two phases, i.e., *design phase* and *runtime phase*, which are shown in Figure 5.29. The design phase is performed off-line to compute and generate the optimized control decisions of both *local* and *global* controllers. Afterwards, these decisions are allocated in a lookup table-based implementation, at design phase, to be used by the global and local controllers at the runtime phase.

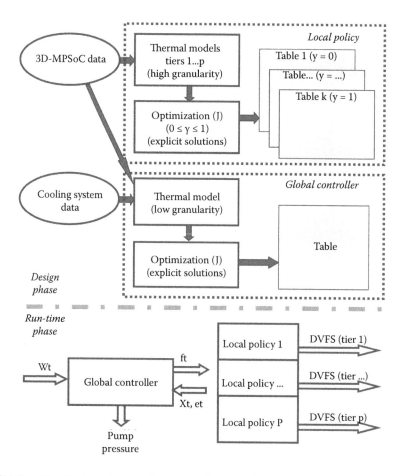

FIGURE 5.29 Design phase and runtime phase of the proposed hierarchical thermal management.

To compute the tables needed for the implementation of the local controllers, all design data related to the structure of the 3D-MPSoC (e.g., elements layout, thermal conductivity of materials, etc.) are used to create an accurate thermal model for each one of the p tiers composing the 3D-MPSoC (upper part of Figure 5.29). This model has a fine granularity and can be formulated as an optimization problem. Different explicit solutions (for various values of the input parameters and optimization goals) are then stored in tables to be used at runtime. To compute the table needed for the implementation of the global controller, we build a coarse-grained thermal model of the 3D-MPSoC and of the cooling system (e.g., the available pumping power values, microchannel layout, etc.).

During runtime, both the global and local controllers apply the rules stored in the aforementioned lookup tables. Each local controller generates the frequency setting for its tier elements, at the processing element-level granularity, while the global controller sets the pressure for the cooling system pump.

The overall system uses software-driven thermal management. That is, the control action is done by software routines (for both the local and global controllers) that access the precomputed data in the tables. These tables represent the control *policies*. Their computation is described in the following subsections.

Since the global controller performs the major decisions by frequent communication with all local controllers and performing the major control actions to them and the coolant system, a dedicated thread resembling the global controller routine is always active. Moreover, this thread is assigned to a single dedicated processing unit that is fully utilized. Being fully utilized implies that this element should be exposed to the maximum cooling ability available in the system. Thus, in our case of 3D-MPSoC with liquid cooling, a processing element that is nearest to the fluid inlet port(s) at any tier is a possible candidate to be allocated to the global control routine [36].

5.4.4 POLICY COMPUTATION: GLOBAL THERMAL CONTROLLER

The *global thermal controller* is the unit responsible for the global joint operation of all local controllers and for the pump control that sets the coolant pressure.

The workload to be dispatched to each local controller is stored in vector $\mathbf{f}_\tau$. The p entries of this vector contain the average frequency of operation at which each local controller has to work in order to execute the workload assigned to its controlled tier by the global unit.

The global controller policy minimizes power and undone workload. Furthermore, the performance requirements coming from the scheduler have to be fulfilled and the maximum temperature constraint satisfied. The problem can be defined as follows (see [50] for details):

$$J = \sum_{\tau=1}^{h} \left(\left\| \mathbf{R}\mathbf{p}_\tau \right\| + \left\| \mathbf{T}\mathbf{u}_\tau \right\| \right) \tag{5.71}$$

$$min\ J \tag{5.72}$$

$$subject\ to\text{: } f_{min} \le \mathbf{f}_\tau \le \mathbf{f}_{max}\ \forall \tau \tag{5.73}$$

$$\mathbf{x}_{\tau+1} = \mathbf{A}\mathbf{x}_\tau + \mathbf{B}\mathbf{p}_\tau \ \forall \ \tau \tag{5.74}$$

$$\tilde{\mathbf{C}}\mathbf{x}_{\tau+1} \preceq \mathbf{t}_{\max} \ \forall \ \tau \tag{5.75}$$

$$\mathbf{u}_\tau \geq \ \forall \ \tau \tag{5.76}$$

$$\mathbf{u}_\tau = \mathbf{w}_\tau - \mathbf{f}_\tau \ \forall \ \tau \tag{5.77}$$

$$\mathbf{l}_\tau \succeq \mu \mathbf{f}_\tau^2 \ \forall \ \tau \tag{5.78}$$

$$-\mathbf{w} \leq \mathbf{m}_{\tau+1} - \mathbf{m}_\tau \leq \mathbf{w} \ \forall \ \tau \tag{5.79}$$

$$0 \leq \mathbf{m}_\tau \leq 1 \ \forall \ \tau \tag{5.80}$$

$$\mathbf{p}_\tau = [\mathbf{l}_\tau; \mathbf{m}_\tau] \ \forall \ \tau \tag{5.81}$$

where matrices **A** and **B** are related to the overall 3D-MPSoC system description. These matrices represent the 3D-MPSoC system using a coarse granularity of the thermal cells, and where the sampling time of the resulting discrete-time system is T_{GC}. We define the horizon of this predictive policy as h [1]. Then, the objective function J is expressed by a sum over the horizon.

In Equation 5.71, the first term, $\|\mathbf{R}\mathbf{p}_\tau\|$, is the norm of the power input vector p weighted by matrix **R**. Power consumption is generated here by two main sources: (1) the workload setting and (2) the liquid cooling pumping power. Vector p is a vector containing normalized power consumption data, the p tiers, and the pumping power. Matrix **R** contains the maximum value of the power consumption of the tiers (first p diagonal entries) and the cooling system (last entry). The second term, $\|\mathbf{T}\mathbf{u}_\tau\|$, is the norm of the required workload, but not yet executed. To this end, the weight matrix **T** quantifies the importance that executing the required workload from the scheduler has in the optimization process. Then, Inequality 5.73 defines a range of working frequencies to be used, but this does not prevent from adding in the optimization problem a limitation on the number of allowed frequency values.

Equation 5.74 defines the evolution of the 3D-MPSoC according to the present state and inputs. Equation 5.75 states that temperature constraints should be respected at all times and in all specified locations. Since the system cannot execute jobs that have not arrived, every entry of $\mathbf{u}_\tau$ has to be greater than or equal to 0, as stated by Equation 5.76. The undone work at time τ, u_τ, is defined by Equation 5.77. Equation 5.78 defines the relation between the power vector **l** and the working frequencies. μ is a technology-dependent constant.

Then, Equations 5.79 and 5.80 define constraints on the liquid cooling management. The normalized pumping power value (**m**) scales, and any time instance τ, from 0 (no liquid injection) to 1 (power at the maximum pressure difference allowable), are shown in Equation 5.80. Moreover, we limit the maximum increment/decrement change in the pumping power value from time (τ) to ($\tau + 1$) by a another normalized value **w**, as shown in Equation 5.79, which models the mechanical dynamics of the pump. Although we assume one pump in the target 3D-MPSoCs, since we use a vector notation for the pumping power and its constraints, our formulation is valid for multiple pumps as well.

Equation 5.81 defines formally the structure of vector **p**. Vector $\mathbf{I} \in \Re^p$ is the power input vector, where p is the number of tiers of 3D-MPSoC.

Finally, we formulate the control problem over an interval of h time steps, which starts at current time τ. Therefore, our approach is predictive. Indeed, the result of the optimization is an optimal sequence of future control moves (i.e., amount of workload to be executed on average for each tier of the 3D-MPSoC, which is stored in vector **f**). Then, we only apply to the target 3D-MPSoC the first samples of such a sequence; the remaining moves are discarded. Thus, at each next time step, a new optimal control problem based on new temperature measurements and required frequencies is solved over a shifted prediction horizon (e.g., the receding-horizon [1] mechanism), which represents a way of transforming an open-loop design methodology into a feedback one, as at every time step the input applied to the process depends on the most recent measurements.

This problem can be transformed so that the solution is given by the linear system

$$\mathbf{y}_{\tau+1} = \mathbf{F}_j \begin{bmatrix} \mathbf{x}_{\tau+1} \\ \mathbf{f}^2\tau \\ \mathbf{w}_\tau^2 \end{bmatrix} + \mathbf{g}_j \qquad (5.82)$$

where **y** is the desired solution as a vector containing the workloads and the pump power, matrix $\mathbf{F}_j$ is a suitable matrix, and $\mathbf{g}_j$ is a suitable vector defined over subregions of the solution space indexed by j. We refer the reader to [1, 49] for details. In [49] an approximate computation method of the regions shows a consistent reduction in the amount of storage space needed with a negligible performance loss.

5.4.5 POLICY COMPUTATION: LOCAL CONTROLLERS

The p local controllers are responsible for the thermal management (e.g., DVFS) of the p tiers of the target 3D-MPSoC. Then, for each tier i the local controller sets frequency and voltage for the $q(i)$ frequency islands (cf. Figure 5.30).

In our hierarchical design, the local controller i receives as input the vector $\mathbf{f}_{t+1}$, which is the average frequency at which island i has to run to execute all the workload assigned to it by the global unit. As second input data, some thermal sensors provide the thermal profile of the 3D-MPSoC island. We assume that the thermal sensors are optimally allocated, as shown in previous work [47]. Thus, the impact of thermal sensor quality and allocation on the management policy is beyond our scope. The local policy computes the frequencies and voltages for all the $q(i)$ units inside island i, as sketched in the dotted box of Figure 5.30. Input data are used as both computing and selection parameters to choose one of the k functions stored in precomputed lookup tables.

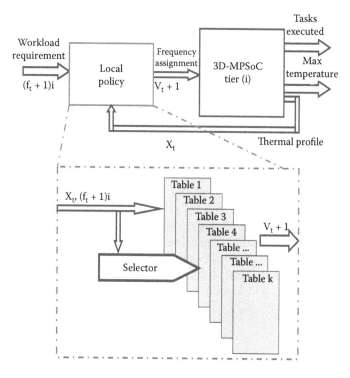

FIGURE 5.30   Local policy controller block diagram.

The local controller decides on the type of optimization to perform—either performance- or power-oriented optimization—and the related policies are stored in the corresponding tables. Specifically, the control policies optimize power and undone workload. We use an optimization parameter γ that weights these two objectives. At the same time, the performance requirement coming from the global controller has to be fulfilled and the maximum temperature constraint satisfied.

The control function is expressed by a policy that is the solution of the following optimization problem (see [50] for more details):

$$J = \sum_{\tau=1}^{h}\left(\left\|\mathbf{Rp}_\tau\right\| + \gamma\left\|\mathbf{Tu}_\tau\right\|\right) \tag{5.83}$$

$$min\ J \tag{5.84}$$

$$subject\ to\text{:}\ 0 \leq v_\tau \leq v_{max}\ \forall\ \tau \tag{5.85}$$

$$\mathbf{x}_{\tau+1} = \mathbf{Ax}_\tau + \mathbf{Bp}_\tau\ \forall\ \tau \tag{5.86}$$

$$\tilde{\mathbf{C}}\mathbf{x}_{\tau+1} \preceq \mathbf{t}_{max}\ \forall\ \tau \tag{5.87}$$

$$\mathbf{u}_\tau \geq \mathbf{0}\ \forall\ \tau \tag{5.88}$$

$$\mathbf{u}_\tau = (\mathbf{f}_\tau)_1 - \sum \mathbf{v}_\tau \ \ \forall \, \tau \tag{5.89}$$

$$\mathbf{p}_\tau \succeq \mu \mathbf{v}_\tau^2 \ \ \forall \, \tau \tag{5.90}$$

where matrices $\mathbf{A}$ and $\mathbf{B}$ are related to the thermal modeling of the specific tier that the local controller is supervising. The objective function J expresses the minimization problem by a weighted sum of two terms, in a similar vein as the global policy is computed, except for the tuning parameter γ. Parameter γ changes according to the specific type of optimization criteria for each tier. It ranges from 0 to 1 in steps of 0.1. We set this parameter at runtime based on the maximum temperature recorded according to a heuristic rule: the hotter the thermal profile, the lower γ is, and vice versa. Thus, the controller performs performance-oriented optimization in the case of a cold thermal profile, but power-saving-oriented optimization in the case of a hot thermal profile. γ is used at runtime to choose from a set of tables, as shown in Figure 5.30.

5.5   SUMMARY

In this chapter, latest improvements in three major fields of thermal management for MPSoCs have been described. Section 5.1 gives a background on state-of-the-art techniques about the fields related to this work. Modeling, algorithms, and system design are the main areas described by this survey.

Section 5.2 introduces mathematical models related to the heat transfer model of the MPSoC, with special emphasis on the way the system cools down and the heat propagates inside the MPSoC. A liquid cooling model of a 3D-MPSoC is also provided. Concepts developed to model the workload of an MPSoC system are also presented. Moreover, some considerations are made about the system energy models. Workload prediction is also introduced, and two estimation techniques are presented.

Section 5.3 introduces techniques to perform a detailed thermal profile estimation of the MPSoC structure. Two techniques are presented here to achieve a temperature estimation by using a few thermal sensors placed in specific locations on the MPSoC.

Section 5.4 describes a distributed thermal management policy suitable for liquid cooling technologies. The policy manages MPSoC working frequencies and microcooling systems to reach their goals in the most effective possible way and consuming the lowest possible amount of resources. The architecture is a distributed structure based on the interaction between a global unit and many small controllers.

REFERENCES

1. A. Bemporad, M. Morari, V. Dua, and E. N. Pistikopoulos. The explicit linear quadratic regulator for constrained systems. In *Automatica*, 2002.
2. B. Black, A. Murali, N. Brekelbaum, J. DeVale, L. Jiang, G. H. Loh, D. McCauley, P. Morrow, D. W. Nelson, D. Pantuso, P. Reed, J. Rupley, S. Shankar, J. Shen, and C. Webb. Die stacking (3D) microarchitecture. In *IEEE MICRO*, 2006.
3. P. Bose. Power-efficient microarchitectural choices at the early design stage. In *PACS*, 2003.

4. T. Boukhobza and F. Hamelin. State and input observability recovering by additional sensor implementation: a graph theoretic approach. In *Automatica*, 2009.

5. S. Boyd and L. Vandenberghe. *Convex Optimization*. Cambridge University Press, Cambridge, 2004.

6. T. Brunschwiler, B. Michel, H. Rothuizen, U. Kloter, B. Wunderle, H. Oppermann, and H. Reichl. Interlayer cooling potential in vertically integrated packages. In *Microsystem Technologies*, 2008.

7. T. Brunschwiler, S. Paredes, U. Drechsler, B. Michel, W. Cesar, G. Toral, Y. Temiz, and Y. Leblebici. Validation of the porous-medium approach to model interlayer-cooled 3D chip stacks. In *3DIC*, 2009.

8. T. Brunschwiler, H. Rothuizen, U. Kloter, H. Reichl, B. Wunderle, H. Oppermann, and B. Michel. Forced convective interlayer cooling potential in vertically integrated packages. In *ITHERM*, 2008.

9. R. L. Burden and J. D. Faires. *Numerical analysis*. Brooks Cole, 2000.

10. K. Choi, R. Soma, and M. Pedram. Dynamic voltage and frequency scaling based on workload decomposition. In *Proceedings of the 2004 International Symposium on Low Power Electronics*, 2004.

11. EMB. WILO MHIE centrifugal pump. http://www.wilo.com/cps/rde/xchg/en/layout.xsl/3707.htm.

12. T. Emi, M. A. Al Faruque, and J. Henkel. Tape: thermal-aware agent-based power economy for multi/many-core architectures. In *ICCAD*, 2009.

13. G. F. Franklin, J. D. Powell, and M. L. Workman. *Digital Control of Dynamic Systems*. McGraw Hill, New York, 2011.

14. J. Haase, M. Damm, D. Hauser, and K. Waldschmidt. Reliability-aware power management of multi-core processors. In *DIPES*, 2006.

15. C. J. Hughes, J. Srinivasan, and S. V. Adve. Saving energy with architectural and frequency adaptations for multimedia applications. In *IEEE Micro*, 2001.

16. W.-L. Hung, G. M. Link, Yuan Xie, N. Vijaykrishnan, and M. J. Irwin. Interconnect and thermal-aware floorplanning for 3D microprocessors. In *ISQED*, 2006.

17. S. Joshi and S. Boyd. Sensor selection via convex optimization. *Transactions on Signal Processing*, 57(2): 451–462, 2009.

18. H. Jung and M. Pedram. Continuous frequency adjustment technique based on dynamic workload prediction. In *VLSI Design*, 2008.

19. P. Kongetira, K. Aingaran, and K. Olukotun. Niagara: a 32-way multithreaded SPARC processor. In *IEEE MICRO*, 2005.

20. Laing. Laing 12 volt DC pumps datasheets. http://www.lainginc.com/pdf/DDC3_LTI_USletter_BR23.pdf.

21. A. J. Laub, M. Heath, C. Paige, and R. Ward. Computation of system balancing transformations and other applications of simultaneous diagonalization algorithms. *IEEE Transactions on Automatic Control*, 32(2): 115–122, 1987.

22. P. P. P. M. Lerou, H. J. M. Ter Brake, H. J. Holland, J. F. Burger, and H. Rogalla. Insight in clogging of MEMS based micro cryogenic coolers. *Applied Physics Letters*, 90, 2007.

23. P. P. P. M. Lerou, G. C. F. Venhorst, C. F. Berends, T. T. Veenstra, M. Blom, J. F. Burger, H. J. M. Ter Brake, and H. Rogalla. Fabrication of a micro cryogenic cooler using MEMS-technology. *Journal of Micromechanics and Microengineering*, 16(10), 2006.

24. S. O. Memik, R. Mukherjee, N. Min, and L. Jieyi. Optimizing thermal sensor allocation for microprocessors. In *IEEE TCAD*, 2008.

25. A. Milenkovic and V. Milutinovic. A quantitative analysis of wiring lengths in 2D and 3D VLSI. In *Microelectronics Journal*, 29(6): 313–321, 1998.

26. R. Mukherjee and S. O. Memik. Physical aware frequency selection for dynamic thermal management in multi-core systems. In *ICCAD*, 2006.

27. S. Murali, A. Mutapcic, D. Atienza, R. Gupta, S. Boyd, L. Benini, and G. De Micheli. Temperature control of high performance multicore platforms using convex optimization. In *DATE*, 2008.

28. A. V. Oppenheim and R. W. Schafer. *Discrete-time signal processing*. Prentice-Hall, Upper Saddle River, NJ, 1989.

29. G. Paci, F. Poletti, and L. Benini. Exploring temperature-aware design in low-power mpsocs. In *DATE*, 2006.

30. A. Papoulis and S. U. Pillai. *Probability, Random Variables and Stochastic Processes*. McGraw Hill, New York, 2002.

31. K. Puttaswamy and G. H. Loh. Thermal analysis of a 3D die-stacked high performance microprocessor. In *GLSVLSI*, 2006.

32. J. M. Rabaey. *Low power design essentials*. Springer, Berlin, 2009.

33. A. Ramalingam, F. Liu, S. R. Nassif, and D. Z. Pan. Accurate thermal analysis considering nonlinear thermal conductivity. In *ISQED 2006*, 2006.

34. R. Rao, S. Vrudhula, C. Chakrabarti, and N. Chang. An optimal analytical solution for processor speed control with thermal constraints. In *Proceedings of the 2006 International Symposium on Low Power Electronics*, 2006.

35. R. C. Chu. Advanced cooling technology for leading-edge computer products. In *5th International Conference on Solid-State and Integrated Circuit Technology*, 1998.

36. M. M. Sabry, A. K. Coskun, and D. Atienza. Fuzzy control for enforcing energy efficiency in high-performance 3D systems. In *ICCAD*, 2010.

37. J. Schtze, H. Ilgen, and W. R. Fahrner. An integrated micro cooling system for electronic circuits. In *IEEE Transactions on Industrial Electronics*, 48(2): 281–285, 2001.

38. O. Semenov, A. Vassighi, and M. Sachdev. Impact of self-heating effect on long-term reliability and performance degradation in CMOS circuits. In *IEEE T-D&M*, 2006.

39. S. Sharifi and T. S. Rosing. An analytical model for the upper bound on temperature differences on a chip. In *GLSVLSI*, 2008.

40. K. Skadron, M. R. Stan, K. Sankaranarayanan, Wei Huang, S. Velusamy, and D. Tarjan. Temperature-aware microarchitecture: modeling and implementation. In *TACO*, 2004.

41. A. Sridhar, A. Vincenzi, M. Ruggiero, T. Brunschwiler, and D. Atienza. 3D-ICE: fast compact transient thermal modeling for 3D-ICS with inter-tier liquid cooling. In *ICCAD*, 2010.

42. A. Sridhar, A. Vincenzi, M. Ruggiero, T. Brunschwiler, and D. Atienza. Compact transient thermal model for 3D ICS with liquid cooling via enhanced heat transfer cavity geometries. In *THERMINIC*, 2010.

43. C. Sumana and C. Venkateswarlu. Optimal selection of sensors for state estimation in a reactive distillation process. *Process Control*, 2009.

44. D. B. Tuckerman and R. F. W Pease. High-performance heat sinking for VLSI. *IEEE Electron Device Letters*, 2(5): 126–126, 1981.

45. K. Waldschmidt. Robustness in SOC design. In *DSD 2006*, 2006.

46. F. Zanini, D. Atienza, A. K. Coskun, and G. De Micheli. Optimal multiprocessor SOC thermal simulation via adaptive differential equation solvers. In *VLSISoC*, 2009.

47. F. Zanini, D. Atienza, C. N. Jones, and G. De Micheli. Temperature sensor placement in thermal management systems for MPSoCs. In *ISCAS*, 2010.

48. F. Zanini, D. Atienza, and G. De Micheli. A combined sensor placement and convex optimization approach for thermal management in 3D-MPSoC with liquid cooling. In *Integration, the VLSI Journal*, 46(1): 33–43, 2013.

49. F. Zanini, David Atienza, G. De Micheli, and S. P. Boyd. Online convex optimization-based algorithm for thermal management of MPSoCs. In *GLSVLSI*, 2010.

50. F. Zanini, M. M. Sabry, D. Atienza, and G. De Micheli. Hierarchical thermal management policy for high-performance 3D systems with liquid cooling. In *IEEE Journal on Emerging and Selected Topics in Circuits and Systems*, 1(2): 88–101, 2011.

6 Emerging Interconnect Technologies for 3D Networks-on-Chip

Rohit Sharma and Kiyoung Choi

CONTENTS

6.1 INTRODUCTION

Historically, microchip design has some or all of these functional blocks: computational, storage, communication, and I/O. Microchip technologies have evolved from large-scale integrated (LSI) to very-large-scale integrated (VLSI) to ultra-large-scale integrated (ULSI) systems. The current ULSI technology, where the chip itself constitutes the entire functional system, defines a modern system-on-chip [1]. As per the International Technology Roadmap for Semiconductors (ITRS) projections, with every next-generation technology node, interconnect effects dominate performance. In that, relative delay in global wires could be manyfold longer than that of local wires or logic gates [2]. Process variations, cross talk, and electromagnetic interference (EMI) can further degrade the performance of these global interconnects. With technology scaling and faster speeds, global synchronization is becoming a mirage. Designers are often required to adopt alternative timing mechanisms including designing globally asynchronous locally synchronous (GALS) chips. While the last decade has seen significantly reduced design cycles, complexity has scaled up several times over the same period of time. Thus, we need a modular approach to design hardware and software, which allows reuse of IPs, so that the key performance metrics, such as reliability, scalability, energy bounds, and manufacturing costs, are met.

Designing today's systems-on-chips (SoCs) that perform functions such as digital signal and graphics processing is a complex exercise [3–5]. Most of these SoCs operate on different clock frequencies, thus becoming distributed systems on a single silicon wafer. This results in a fully distributed communication pattern with no global control. Communication structure in typical SoCs, such as single-chip embedded systems, mobile phones, and HDTVs, could be anything from conventional bus-based to dedicated point-to-point links to ad hoc, irregular networks [1]. However, there are multiple concerns with bus-based communication structures, such as parasitic effects, timing control, limited bandwidth, and arbitration delay [6, 7]. Networks, on the other hand, can be preferred over buses because of higher bandwidth, pipelining, and multiple concurrent communication support. Also, dedicated point-to-point links may not be a viable solution, as the number of links may increase much faster with increasing number of cores. For many core systems with much less design cycle time, a shared and segmented global communication structure is essential. This in turn would mean that we come up with a segmented SoC communication structure with shorter wires for better signal integrity and multiplexed buses for increased throughput and lower energy budget. A network-on-chip (NoC) can be seen as a promising solution for optimum SoC design by integration of many cores that provides answers to some of the above-mentioned design challenges [1, 3].

An NoC is a reconfigurable interconnection of processors, distributed storage elements, and I/Os that are connected using routers or switches. The layered, reconfigurable network aids in efficient communication between these elements using data packets and provides a plug-and-play use of various components. The NoC exploits the basic methods and tools used in general computer networks and guarantees highly reliable and robust communication [1–7]. Figure 6.1 gives a simple illustration of an $n \times n$ mesh-type NoC with its key components, while Figure 6.2 gives a typical structure of a data packet.

There are three main components in an NoC: network adapters, routing nodes, and the interconnect links [1]. The network adapters act as interfaces between the computational units and the communication channel. Routing strategies based on specific protocols are implemented using the routing nodes. The interconnect links

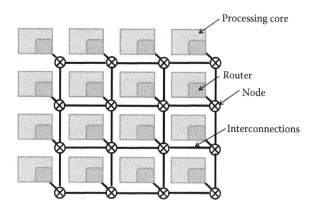

FIGURE 6.1   Schematic of a two-dimensional $n \times n$ mesh NoC.

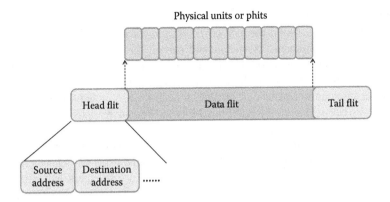

FIGURE 6.2 Typical packet structure.

act as the physical layer and consist of many communication channels. The preferred communication mechanism to transmit data in NoCs is packet switching. NoCs use energy-efficient routing algorithms and tables for optimum packet transmission using routers. Today, NoC designers aspire for the optimum energy-delay-area metric for these routers. The interconnect links (mostly copper wires) are the heart of the physical layer in the NoC. Throughput and energy budget in an NoC are primarily determined by these interconnect links. While metal interconnects are by far the most common choice, emerging interconnect technologies like nanophotonics, wireless, optical, and carbon-based interconnects are extensively researched as other alternatives. We shall be discussing the pros and cons of these interconnect technologies in detail in Section 6.3.

The performance and reliability of the network, and thus that of the system, greatly depend on how the network nodes are connected in the physical space, more commonly known as the topology of the network. The topology determines the footprint of the NoC, interconnect layout, and size of the router. Broadly, NoC topologies are classified into two types: the *regular topologies* that offer superior scalability (for example, mesh topology) and the *irregular topologies* that offer the reverse characteristics (for example, star topology) [8]. Dally [9] provided the basic framework of a grid-type *k-ary n-cube* network topology. Here, k is the degree (number of nodes) of each dimension, while n is the number of the dimensions (for example, 2D, 3D). Thus, what we see in Figure 6.1 represents a 4-ary 2D *mesh* topology. Another commonly used regular topology is the *torus* topology, which differs from mesh topology in terms of the type of interconnect links (ring connections) employed to construct the network. Torus and mesh topologies refer to *direct networks*, where at least one core is attached to each node. However, there are examples of *indirect regular networks*, such as the tree topology (binary tree, fat tree, and butterfly tree), that offer hardware efficiency when compared to direct networks [6, 10]. Irregular topologies, on the other hand, have an asymmetric network that scales nonlinearly with respect to energy budget and area requirements [10]. These irregular topologies are more suited for application-specific SoCs that are heterogeneous in nature, with varying sizes, functionality, and communication requirements of the computational elements (cores).

Summarizing this section, it would be logical to highlight some of the key advantages of NoCs. NoCs can use *GALS architecture*, thereby avoiding complex global timing requirements. As mentioned earlier, the router efficiently decouples the computational elements (processor cores) with the communication framework. Compared to buses, NoCs use shorter wires that lead to lower energy consumption and lower interconnect parasitics. Further, it is simpler to model shorter wires with reduced *design complexities*. Reconfigurability in NoCs increases *design productivity* on the one hand, while reducing *design cycle time* on the other hand. A reconfigurable network helps the designers to add/remove elements based on requirements, thus providing highly *scalable architecture*. Wires can be shared using packet switching that results in higher throughput than conventional buses. Over the years, NoC simulation tools have evolved considerably, resulting in lower manufacturing costs. Finally, *testability* is better in NoCs, as they employ efficient error correction schemes.

6.2   3D INTEGRATION AND 3D NETWORKS-ON-CHIP

Continuous devices scaling and performance constraints of global interconnects have led to stacking of multiple dies of integrated circuits (ICs). This vertical stacking has given system designers a window of opportunity to incorporate more functionality into a single package. In this section, we focus on the evolution and advantages of 3D integrated technology and its amalgamation with NoCs to design 3D NoC architectures. Further, we highlight the major design constraints that plague the performance of 3D NoCs.

6.2.1   3D Integrated Circuits

3D ICs provide an answer to the limitations set by long interconnects through stacking active silicon layers. These vertically stacked layers lead to reduction in size and number of the global interconnects, offering an opportunity to meet Moore's law. Therefore, one can expect a significant increase in performance and decrease in power consumption and area with possible integration of CMOS circuits with other technologies [11–13]. 3D ICs offer a number of advantages compared to 2D ICs. These include:

- Shorter global interconnects
- Superior performance
- Lower power consumption
- Higher packing density
- Smaller area (footprint)
- Scope of mixed-technology ICs

However, 3D ICs have significant concerns in the form of thermal considerations. While the overall power dissipation in 3D ICs may be lower due to shorter and lesser global interconnects, the power density is much higher due to the vertically stacked silicon layers. Thus, efficient thermal management is the key to guarantee the performance improvements offered by 3D ICs. For greater understanding on this topic, readers are referred to the various thermal management techniques, such as physical design optimization, use of thermal vias, and microfluidic cooling of the vertical stack reported in the literature [14–16].

Vertical integration of chips has also resulted in a paradigm shift in the way we investigate the interconnect technologies by using vertical interconnects for interstrata communication. The vertical interconnect technologies include microbumps, wire bonding, wireless interconnects using capacitive/inductive coupling, and through-silicon vias (TSVs), of which TSVs offer very-high-density vertical interconnects and are by far the most promising technology [13].

6.2.2 3D Networks-on-Chip

By implementation of vertical integration in on-chip networks, one can design 3D NoCs that outperform their 2D counterparts. Figure 6.3 gives a simple illustration highlighting the merger of these two approaches.

3D NoC architectures are generally classified into two types: *symmetric* and *bus hybrid*. However, the latter lacks concurrent communication in the vertical stack and suffers from possible contention and blocking issues in the vertical intercon-nects. The key performance metrics in 3D NoCs include zero-load latency and power consumption of the network. To optimize these two metrics, the authors in [11] have proposed various 3D NoC topologies, which are shown in Figure 6.4.

The 3D IC, 3D NoC topology, as shown in Figure 6.4d, consists of processing elements (PEs) that are integrated over multiple vertical planes and routers, each of which connects two additional neighboring routers (other than the four adjacent in the same plane) located on the adjacent vertical planes. Such a 3D NoC topology significantly minimizes the zero-load latency as well as the power consumption [11]. However, it is important to note that the worst-case performance may not improve by mere transformation of 2D NoC to 3D NoC. This is due to the fact that the worst-case delay is more sensitive to bandwidths offered by the vertical interconnects and

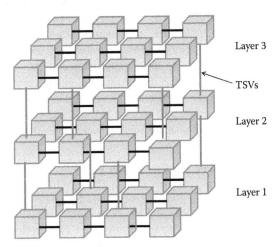

FIGURE 6.3   Vertical integration of *n* × *n* 2D mesh results in a symmetric 3D mesh NoC.

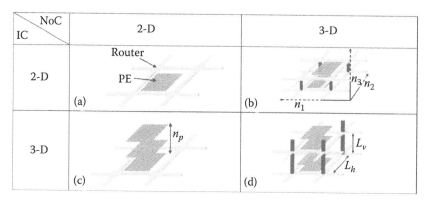

FIGURE 6.4 Various NoC topologies: (a) 2D IC, 2D NoC; (b) 2D IC, 3D NoC; (c) 3D IC, 2D NoC; (d) 3D IC, 3D NoC. (From V. F. Pavlidis and E. G. Friedman, 3-D topologies for networks-on-chip. *IEEE Transactions on Very Large Scale Integration (VLSI) Systems,* 15(10): 1081–1090, 2007. Copyright © 2007 IEEE. Reprinted with permission.)

the size of the network [17, 18]. Additionally, 3D architectures require modified design of the router. One such design is presented by Kim et al., where they propose a dimensionally decomposed (DIMDE) router architecture [19]. This 3D router has benefits of path diversity and higher bandwidth by supporting two vertical interconnects and offers a seamless traffic flow in the XYZ directions.

The physical layer of a 3D NoC architecture consists of longer horizontal interconnects that connect the adjacent nodes in the same layer and shorter vertical interconnects that connect the nodes on different layers. Thus, we have a wiring layout that moves in the XYZ direction with vertical interconnects having lower delay than horizontal interconnects. The basic technology for these vertical and horizontal interconnects is the use of Cu metal lines for intralayer communication and TSVs for interlayer communication. In the deep submicron (DSM) and ultra-DSM era, even with efficient PE and router designs, wiring constraints would be the most important performance bottlenecks in 3D NoCs. The conventional Cu/dielectric interconnect systems have a limited lifetime as we approach future technology nodes. However, there are several other promising technologies that offer alternatives to these metal interconnects. The ITRS predicts development of novel materials to meet the ever-increasing performance requirements of these 3D NoCs. There has been a concerted effort in this direction, and several interesting interconnect technologies are proposed for 3D NoC architectures. In the remainder of this chapter, we shall focus on some of these technological options.

6.3 INTERCONNECT TECHNOLOGIES FOR 3D NETWORKS-ON-CHIP

In this section, we present a detailed overview of the common interconnect technologies used in 3D NoCs, namely, radio frequency (RF)/TSV, optical and photonics, carbon-based, and wireless communication. In that, our focus will be on the relative advantages offered and challenges posed by these competing technologies.

6.3.1   RF/TSV-Based Interconnect Technology for 3D NoCs

Among the various interconnect technologies listed above, the use of TSVs is the most promising one and remains the focus of the majority of 3D integration R&D activities. TSVs in 3D NoCs can be used to communicate with the interlayer cores, providing a vertical communication pathway. Compared to other 3D assembly technologies [20], such as wire bonding, metal bumps, and contactless (wireless) coupling, TSVs offer short, low-loss electrical links with a lower footprint and high density. Use of TSVs provides functional benefits, like reduction in the number and length of global wires and integration of disparate technologies. However, the fabrication process of TSVs is a complex and expensive task and mainly governs the integration density and cost [21]. Some of the most widely used TSV fabrication processes include via first (VF), via middle (VM), and via last (VL) processes as shown in Table 6.1.

3D NoC architectures using TSVs can be fabricated using either the monolithic approach or the stacking approach. In the monolithic approach, front-end processing to fabricate the active device layer is repeated on a single wafer. The backend processing takes care of fabricating the interconnects/TSVs. In the second approach, the individual active device layers are stacked over one another using face-to-face or face-to-back bonding. 3D NoC designs using TSVs show that there is a trade-off between the performance and manufacturing cost. For better thermal management, designers use efficient floor planning where the processors are on one layer below the heat sink, while other components (caches, etc.) are on the other layer, as shown in Figure 6.5. For optimum performance of 3D NoCs, full connectivity between interlayer cores is desired. However, as the number of tiles/nodes increases, there is also an increase in the chip area and manufacturing cost. For example, one would require several hundred TSVs for full layer-layer connectivity of a 4 × 4 mesh-type 3D NoC. If quarter or half layer-layer connectivity is used, the average number of hops increases, thereby severely degrading the network latency and causing communication deadlock [21].

There are several electrical, thermal, and mechanical design issues associated with the use of TSVs in 3D NoCs. From the electrical side, interlayer vertical interconnections require the design of new router architecture to take into account the additional vertical TSV pathways. The authors in [19] have presented a 3D dimensionally decomposed (DimDe) router that supports two additional vertical TSVs and

TABLE 6.1
Comparison between Various Via Processes

	VF	VM	VL
Fabrication step	TSVs fabricated before CMOS process	TSVs fabricated before BEOL, after CMOS	TSVs fabricated after BEOL process
Diameter	<5 μm	~5 μm	>5 μm
Density	High	Medium	Low
Cost	High	Medium	Low

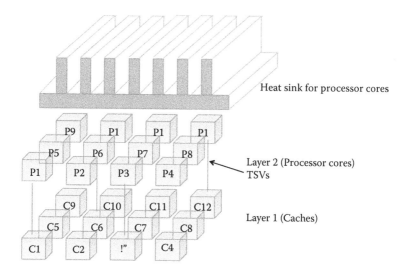

FIGURE 6.5 Simple illustration of a 3D NoC with separate processor and cache layers. The processor layer is placed below the heat sink.

optimizes the cost of 3D NoC switches. This modified router uses a partial crossbar switch that supports TSVs and enables concurrent communication between the different layers of the 3D NoC. Compared to a full 3D crossbar, there is significant reduction in area, power budget, and logic complexity. The DimDe architecture clearly scores over other architectures and can be used for superior electrical performance of 3D NoCs. Increased power density per unit area in 3D NoCs results in the heat dissipation of an inner active layer. Routers in 3D NoCs can become a thermal hotspot due to excessive packet transmission and heat generated due to 3D stacking. Thermally aware router architecture design is an important research topic.

It is also important to consider the various thermomechanical issues in 3D NoCs due to the use of TSVs. TSVs are seen to induce a built-in stress that may result in mechanical failures (e.g., delamination, peel, fatigue, etc.) and electrical performance degradation (e.g., parameter shifts, increased variability, EM, etc.). *Misalignment* during the wafer bonding process and *random open defects* due to thermal compression should be minimized to guarantee superior TSV yield [22].

While TSVs can provide vertical connection between interlayer cores, high-speed interconnect shortcuts have been proposed for communication between cores on the same layer. These shortcuts can be designed using multiband RF interconnects. The concept of RF interconnects, as explained by the authors in [23], is based on transmission of waves, rather than voltage signaling. When compared to conventional transmission line interconnects, RF interconnects achieved significantly lower latency and energy consumption. Thus, one can use RF interconnects overlaid on mesh architecture for intralayer and TSVs for interlayer communication that provides robust and reliable interconnect architecture for 3D NoCs. Although there are several design challenges that need to be addressed, by far, the use of RF/TSV-based interconnect architecture seems to be a promising option for 3D NoCs.

6.3.2 Optical/Photonics-Based Hybrid 3D NoCs

In this section, we discuss the use of optical and photonics-based interconnect technology for intra- and interlayer communication in 3D NoCs. The use of optics or optical interconnects for on-chip and chip-chip communication has been a topic of extensive research in the last decade. There has been seminal work that has been reported by authors on the potential benefits of communication using optical media [24, 25]. With continuous scaling in interconnect technology, conventional copper-based interconnects suffer from signal and clock integrity issues. Optical technology promises to address several physical problems of metal interconnects, including precise clock distribution, system synchronization, bandwidth and density of long interconnections, and reduction of power dissipation. Further, optical interconnects overcome a broad range of design challenges commonly encountered in metallic interconnects, including cross talk, voltage isolation, wave reflection, impedance matching, and pin inductance [25]. While electrical-optical translation costs, CMOS incompatibility, and integration issues were major hindrances at the onset, rapid progress in CMOS-compatible detectors, modulators, and light sources has overcome the initial skepticism about optoelectronic technology [26].

Figure 6.6 gives a simplified schematic of an on-chip optoelectronic interconnection system with associated components, which comprises three major components: a transmitter, a waveguide, and a receiver. The transmitter consists of a laser source, a modulator, and a driver circuit. The laser source provides light to the modulator that converts electrical signals into a modulated optical signal. Waveguides are the optical links (interconnects) through which light gets transmitted. Silicon and polymer waveguides are widely used optical links. The optical receiver performs the reverse optical-to-electrical conversion from light to electrical data. It consists of a photodetector, an amplifier stage, and an additional filtering stage for wave division multiplexing.

The situation, however, is different if one intends to connect processor cores in a chip multiprocessor (CMP) using an optical network. Optical bus architecture for communication in a CMP was first proposed by the authors in [26] with a loop-shaped bus made of optical waveguides (residing on a dedicated Si layer), as shown in Figure 6.7. There are multiple nodes (or switches) that are connected to the bus, which provide an

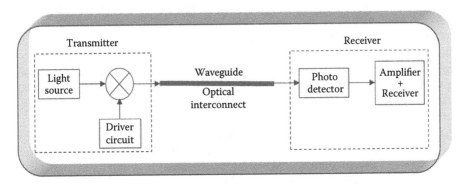

FIGURE 6.6   Illustration of an on-chip optical interconnect architecture.

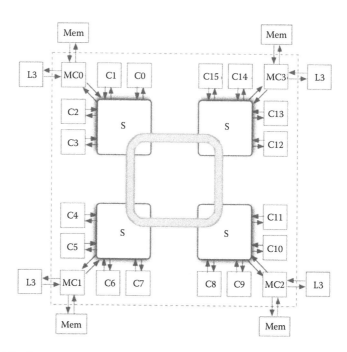

FIGURE 6.7   Proposed CMP floor plan with an overlaid optical bus. (From N. Kirman et al., Leveraging optical technology in future bus-based chip multiprocessors, In *Proceedings of IEEE/ACM International Symposium on Microarchitecture*, 2006, pp. 492–503. Copyright © 2006 IEEE. Reprinted with permission.)

optoelectronic interface between the optical bus and the electrical processor core(s). In this interconnect scheme, the optical loop constitutes the top level of the hierarchy, while the nodes deliver information to processors via electrical sublevels. Such a communication scheme requires area- and energy-aware bus topologies. The area budget includes estimates for the active, optical, and metal layers for all possible topologies. In the active layer area, the proposed design accounts for electrical switches in each node, as well as transmitters and receivers on the optical bus. The area occupied in the optical layer is calculated as the sum of the waveguide, modulator, detector, and wave-selective filter areas. The power consumption of the interconnect system is categorized into two components: the power consumed in the electrical sublevels (switches and wiring) and the power consumed in the optical source. It is seen that the area requirement of the optical layer is the highest, while the power budget is the lowest for all the given topologies. The authors conclude that the four-node configuration is preferable, as the power consumption of the optical components is relatively low compared to that of the electrical subnetwork.

A true 3D optical NoC (ONoC) combines the advantage of optical and 3D technologies providing low latency and high bandwidth with significantly lower power consumption [27–31]. Typically, 3D ONoC is organized in a multilayer configuration with an electronic layer consisting of heterogeneous processing elements and routers. One or more layers stacked over this electronic layer provides most of the memory storage. The topmost layer is the communication layer that consists of the optical components. Such a hybrid architecture combines the advantages of photonic NoC with 3D integration, where the electronic networks handle smaller packets (data and control), while the optical network handles larger data packets, ensuring lower energy dissipation along with low loss in long optical waveguides. One of the most important components of an ONoC is the design of optical routers with a switching fabric that implements routing and flow control functions at its core. These switching elements can be built using a microresonator. Typically, an $n \times n$ optical crossbar requires n^2 microresonators and $2n$ crossing waveguides. Such a crossbar for 3D ONoC has been proposed by the authors in [27, 28] and consists of a data information processing unit (DIPU) and a control information processing unit (CIPU). Typically, the DIPU is the optical domain, while the CIPU is the traditional CMOS-based electronic domain. As compared to traditional 3D crossbars, we can obtain nearly two-thirds loss reduction by using these 3D optical crossbars. Figure 6.8 gives the schematic of a 3D mesh topology with an overlaid optical network.

For optimum utilization of the optical channel, a connection-oriented communication protocol with a dimension order routing algorithm is preferred. Since optical buffers are not available, the communication protocol ensures that no buffers are required for the optical data. In the above architecture, the routing algorithm and the optimized crossbar design ensure lower delay and better throughput than its 2D counterpart. Network topology plays an important role in the performance of the ONoC. The authors in [30–31] have proposed several wavelength-routed ONoC topologies considering the properties of the optical links as well as their placement constraints. In that, the optical ring-based topology is simpler to design with acceptable power dissipation. However, this topology is limited to target systems with smaller die sizes with simple connectivity requirements.

One major problem with ONoCs is the need for longer waveguides that may result in power loss and back-reflections. Recently, the use of nanophotonic interconnects (NIs) is envisaged for reliable communication between future multicore systems. The on-chip multilayer photonic (OCMP) NoC architecture is an NI-based interconnect architecture that consists of 16 decomposed NI-based crossbars placed on four

Overlaid optical fabric

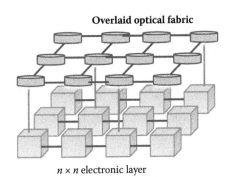

$n \times n$ electronic layer

FIGURE 6.8   A 4 × 4 3D architecture with distinct optical and electronic layers. The optical fabric is laid over the electronic layer.

optical communication layers as proposed in [32]. The OCMP architecture removes optical waveguide crossings as well as the use of meandering waveguide placement, thereby reducing optical power losses even further. The NI-based 3D NoC seems to be a promising candidate for future many-core and multicore architectures.

6.3.3   3D Wireless NoCs

Wireless communication between interlayer cores in 3D NoCs is quite a radical idea that aims to achieve higher throughput and better latency with lower fabrication and testing costs than those using the conventional TSVs or optical communication. The idea here is to design a hybrid architecture that consists of a conventional mesh-based wired topology for communication within individual layers and a vertical wireless channel for interlayer communication. While this concept seems to be a promising alternative, readers must note that it is still in its nascent developmental stage.

   In the past, the concept of wireless communication in planar 2D NoCs has been proposed by several researchers. A wireless channel in 2D NoCs works on high-bandwidth, single-hop, long-range communication, as against multihop communication in regular wired NoCs, resulting in lower latency, lower power consumption, and easier routing schemes [29]. The concept of wireless communication in 2D NoCs was first highlighted by Floyd et al. in [33], where a clock distribution network was implemented using wireless interconnects. Also, wireless 2D NoCs using an ultrawideband communication scheme have been proposed in [34]. Typically, a wireless channel in a planar NoC has an antenna, network architecture, and transceiver circuits as its basic organizational elements. The idea here is to divide the network into multiple subnets with wired intra-subnet communication and wireless links communicating between the subnets, as shown in Figure 6.9. Each subnet consists of a base station for setting up the wireless link. Unlike traditional wired NoCs, individual subnets in a wireless NoC (WiNoC) can have different architectures, thereby having a heterogeneous design that results in significantly improved latency and throughput.

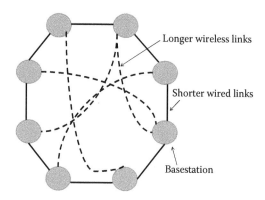

FIGURE 6.9   Hybrid wireless 2D NoC with subnets communicating using wireless links.

The wireless links are established using mm-wave metallic antennas and infrared carbon nanotube (IR CNT) antennas. Wireless communication for shorter distances (typically 1 mm) may not be more efficient than wired lines. A greater number of wireless nodes will result in area penalty due to several antennas and their associated transceiver circuit. For longer-range wireless communication, metal antennas with zigzag patterns can be used to achieve long-distance (over several millimeters) communication with operating frequencies in tens of GHz [35]. However, careful selection of antenna configuration and dielectric material can increase the operating frequency of the wireless channel in the range of 50–100 GHz. One major issue with mm-wave metal antennas is their area overhead (typically 1–2 mm). In this scenario, nanoscale antennas based on CNTs can provide THz/optical frequency ranges with significantly smaller sizes [36]. The problem, however, is the fact that traditional fabrication techniques are incompatible for overcoming the manufacturing challenges in CNT-based optical antennas. Even if we expect better control over CNT growth in the coming days, there are several issues associated with this type of communication scheme with respect to establishing interlayer channels for 3D NoCs. The antenna-based communication scheme discussed above propagates surface waves that allow communication between subnets in a 2D architecture. For 3D wireless channels between IPs stacked over each other, we need alternative wireless technologies. One such approach is the use of inductive coupling for vertical communication between interlayer IPs.

Inductive coupling has been extensively researched for applications in 3D ICs [37–44]. Davis et al. [37] have used inductive coupling for Fast Fourier Transform (FFT) applications and benchmarked it against other technologies for area and power budget. In [38], the authors have demonstrated a 1 Gbps link between two dies with multiple inductive coupling links. The proposed link outperforms other wireless communication techniques. They further extend the work to obtain 1 Tbps between two layers with minimized cross talk effects [39]. In [40], the authors proposed a bidirectional communication channel for vertical links. The area and power requirements using inductive couplers can be further reduced using burst communication, as shown in [41]. In [42], Choi et al. have proposed chip-to-chip communication using integrator circuits. Wireless communication using inductive coupling has several advantages. The transceiver circuit, including the inductors, can be implemented in a standard CMOS process that is very cost-effective. Tests of individual dies can be performed before they are stacked over each other. Finally, unlike TSV techniques, inductive coupling does not cause any mechanical stresses [37].

The concept of 3D wireless NoC using inductive coupling was first demonstrated by Lee et al. [43]. The proposed architecture assumes a 4 × 4 × 3 3D mesh network, as shown in Figure 6.10a. We have used an integration-based transceiver circuit for inductive coupling. The intralayer links are 32 bits wide, and the routing method is dimensional wormhole routing. Individual routers have eight data transceivers and a single clock transceiver, as shown in Figure 6.10b. The transceiver has separate coils for data and clock transmission with a resultant vertical link of datawidth of 8 bits. The total area of the six-port router in 90 nm technology is 0.13 mm^2, which includes an area budget of 0.03 mm^2 for the nine coils.

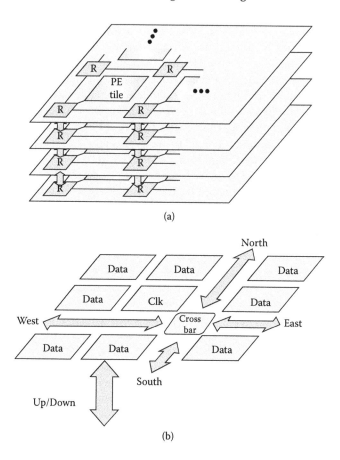

FIGURE 6.10 System architecture for 3D wireless NoC using inductive coupling: (a) mesh network and (b) router layout for inductive coupling.

Considering the characteristics of inductive coupling, we proposed a token bus protocol and a global clock transmission scheme. In our proposed approach, only the bottom layer generates a clock and broadcasts it to all layers, including the bottom layer itself. Instead of transmitting a data pulse at the rising edge of its local clock, the transmitter sends out the pulse at the rising edge of the broadcast clock. All the receivers generate a sense signal from the same clock received from the bottom layer. Due to the layer-by-layer difference of the timing for the sensing, a delay adjustment circuit is needed. However, since the integration receiver is used, there is a relatively large timing margin, and thus the delay adjustment need not be so precise. For the proposed wireless communication, we decided to use the token bus protocol. As can be figured out from the name, initially one layer has the token and the token holder gets the right to transmit. When a layer is done with transmitting data, or if it has nothing to send, the next layer gets the token. There can be many variations on when to pass the token or how to determine the next token holder. We decided to use a simple method, as follows. Each token holder sends one full packet, and then broadcasts an announcement such that the upper layer (wraps around from the top to the bottom) gets the token. We can solve the valid signal problem using the header flit containing type, source, destination, and packet size information. If the type indicates that the packet is just for an announcement, other layers will know that the transmitter layer has nothing to send and the token will be passed. Our proposed scheme obtained aggregate throughput of 4.7 flits per cycle under uniform random traffic, which was 78% higher than that for the naive time division scheme. The proposed scheme also outperformed the time division scheme in terms of latency. Overall, this scheme looks promising for wireless communication in 3D NoCs. However, one needs to investigate several issues associated with its implementation and its concurrent use in 2D communication over the same network.

6.4 SUMMARY

This chapter proposes the various interconnect technologies that can be used for communication between vertically stacked IPs in 3D NoCs. Overall, we have seen that three major technologies are potential candidates. In the traditional copper-based wired domain, the most widely used technology is TSVs that form vertically wired channels. However, TSVs suffer from thermal and mechanical issues that need to be addressed. In the optical technology domain, power loss can be a major design issue that can be addressed by employing NI-based interconnect channels. Finally, inductive coupling-based wireless channels for vertical communication can be another alternative technology. The authors feel that an optimum 3D NoC interconnect scheme should have a mix of some or all of these disparate technologies. One way to look at this would be a true hybrid architecture where the planar interconnections are the traditional transmission lines, while the vertical interconnects could be one of the above interconnect technologies. For the moment, this is an open research topic that needs further investigations to answer this question.

ACKNOWLEDGMENTS

The authors gratefully acknowledge the technical help and support received from the members of the Design Automation Lab at Seoul National University. The authors also thank Dr. Noriyuki Miura of the Kuroda Lab at Keio University for the deep technical insight provided on wireless communication in 3D ICs using through-chip interface technology.

REFERENCES

1. T. Bjerregaard and S. Mahadevan, A survey of research and practices of network-on-chip, *ACM Computing Surveys*, 38(1): 1–51, 2006.
2. International Technology Roadmap for Semiconductors, 2011, www.itrs.net.
3. D. Geer, Networks on processors improve on-chip communications, *Computer*, 42(3): 17–20, 2009.
4. L. Benini and G. De Micheli, Networks on chips: a new SoC paradigm, *Computer*, 35(1): 70–78, 2002.

5. W. J. Dally and B. Towles, Route packets, not wires: on-chip interconnection networks, In *Proceedings of Design Automation Conference*, 2001, pp. 684–689.
6. P. Guerrier and A. Greiner, A generic architecture for on-chip packet-switched interconnections, In *Proceedings of Design, Automation and Test in Europe Conference and Exhibition*, 2000, pp. 250–256.
7. J. Cong, An interconnect-centric design flow for nanometer technologies, In *Proceedings of the IEEE*, 89(4): 505–528, 2001.
8. T. Sharma, *Fault tolerant network on chips topologies*, University of Stuttgart, 2009.
9. W. J. Dally, Performance analysis of k-ary n-cube interconnection networks, *IEEE Transactions on Computers*, 39(6): 775–785, 1990.
10. P. P. Pande, C. Grecu, M. Jones, A. Ivanov, and R. Saleh, Effect of traffic localization on energy dissipation in NoC-based interconnect, In *IEEE International Symposium on Circuits and Systems*, 2005, vol. 2, pp. 1774–1777.
11. V. F. Pavlidis and E. G. Friedman, 3-D topologies for networks-on-chip, *IEEE Transactions on Very Large Scale Integration (VLSI) Systems*, 15(10): 1081–1090, 2007.
12. R. J. Gutmann, J.-Q. Lu, Y. Kwon, J. F. McDonald, and T. S. Cale, Three-dimensional (3D) ICs: a technology platform for integrated systems and opportunities for new polymeric adhesives, In *First International IEEE Conference on Polymers and Adhesives in Microelectronics and Photonics*, 2001, pp. 173–180.
13. W. R. Davis, J. Wilson, S. Mick, J. Xu, H. Hao, C. Mineo, A. M. Sule, M. Steer, and P. D. Franzon, Demystifying 3D ICs: the pros and cons of going vertical, *IEEE Design and Test of Computers*, 22(6): 498–510, 2005.
14. B. Goplen and S. Sapatnekar, Efficient thermal placement of standard cells in 3D ICs using a force directed approach, In *International Conference on Computer Aided Design*, 2003, pp. 86–89.
15. J. Cong and Y. Zhang, Thermal via planning for 3-D ICs, In *IEEE/ACM International Conference on Computer-Aided Design*, 2005, pp. 745–752.
16. B. Dang, P. Joseph, M. Bakir, T. Spencer, P. Kohl, and J. Meindl, Wafer-level microfluidic cooling interconnects for GSI, In *Proceedings of the IEEE Interconnect Technology Conference*, 2005, pp. 180–182.
17. A.-M. Rahmani, K. R. Vaddina, K. Latif, P. Liljeberg, J. Plosila, and H. Tenhunen, Generic monitoring and management infrastructure for 3D NoC-bus hybrid architectures, In *IEEE/ACM International Symposium on Networks on Chip (NoCS)*, 2012, pp. 177–184.
18. Y. Qian, Z. Lu, and W. Dou, From 2D to 3D NoCs: a case study on worst-case communication performance, In *IEEE/ACM International Conference on Computer-Aided Design—Digest of Technical Papers*, 2009, pp. 555–562.
19. J. Kim, et al., A novel dimensionally-decomposed router for on-chip communication in 3D architectures, In *Proceedings of International Symposium on Computer Architecture*, 2007, pp. 138–149.
20. K. Salah, A. El-Rouby, H. Ragai, and Y. Ismail, 3D/TSV enabling technologies for SOC/NOC: modeling and design challenges, in *International Conference on Microelectronics*, 2010, pp. 268–271.
21. T. C. Xu, P. Liljeberg, and H. Tenhunen, A study of through silicon via impact to 3D network-on-chip design, In *International Conference on Electronics and Information Engineering*, 2010, vol. 1, pp. 333–337.
22. I. Loi, S. Mitra, T. H. Lee, S. Fujita, and L. Benini, A low-overhead fault tolerance scheme for TSV-based 3D network on chip links, In *IEEE/ACM International Conference on Computer-Aided Design*, 2008, pp. 598–602.
23. M. F. Chang, J. Cong, A. Kaplan, M. Naik, G. Reinman, E. Socher, and S.-W. Tam, CMP network-on-chip overlaid with multi-band RF-interconnect, In *IEEE 14th International Symposium on High Performance Computer Architecture*, 2008, pp. 191–202.

24. J. W. Goodman, F. J. Leonberger, S.-Y. Kung, and R. A. Athale, Optical interconnections for VLSI systems, *Proceedings of the IEEE*, 72(7): 850–866, 1984.

25. D. A. B. Miller, Rationale and challenges for optical interconnects to electronic chips, *Proceedings of the IEEE*, 88(6): 728–749, 2000.

26. N. Kirman, et al., Leveraging optical technology in future bus-based chip multiprocessors, In *Proceedings of IEEE/ACM International Symposium on Microarchitecture*, 2006, pp. 492–503.

27. H. Gu and J. Xu, Design of 3D optical network on chip, In *Symposium on Photonics and Optoelectronics*, 2009, pp. 1–4.

28. Y. Yaoyao, D. Lian, J. Xu, J. Ouyang, M. K. Hung, and Y. Xie, 3D optical networks-on-chip (NoC) for multiprocessor systems-on-chip (MPSoC), In *IEEE International Conference on 3D System Integration*, 2009, pp. 1–6.

29. L. P. Carloni, P. Pande, and Yuan Xie, Networks-on-chip in emerging interconnect paradigms: advantages and challenges, In *3rd ACM/IEEE International Symposium on Networks-on-Chip*, 2009, pp. 93–102.

30. L. Ramini, P. Grani, S. Bartolini, and D. Bertozzi, Contrasting wavelength-routed optical NoC topologies for power-efficient 3D-stacked multicore processors using physical-layer analysis, In *Design, Automation and Test in Europe Conference and Exhibition*, 2013, pp. 1589–1594.

31. L. Ramini and D. Bertozzi, Power efficiency of wavelength-routed optical NoC topologies for global connectivity of 3D multi-core processors, In *Proceedings of Fifth International Workshop on Network on Chip Architectures*, 2012, pp. 25–30.

32. R. W. Morris, A. K. Kodi, A. Louri, and R. D. Whaley, Three-dimensional stacked nanophotonic network-on-chip architecture with minimal reconfiguration, *IEEE Transactions on Computers*, 63(1): 243–255, 2014.

33. B. A. Floyd, C.-M. Hung, and K. K. O, Intra-chip wireless interconnect for clock distribution implemented with integrated antennas, receivers and transmitters, *IEEE Journal of Solid-State Circuits*, 37(5): 543–552, 2002.

34. D. Zhao and Y. Wang, MTNet: design of a wireless test framework for heterogeneous nanometer systems-on-chip, *IEEE Transactions on Very Large Scale Integration (VLSI) Systems*, 16(8): 1046–1057, 2008.

35. L. Jau, H.-T. Wu, Y. Su, L. Gao, A. Sugavanam, J. E. Brewer, and K. K. O, Communication using antennas fabricated in silicon integrated circuits, *IEEE Journal of Solid-State Circuits*, 42(8): 1678–1687, 2007.

36. J. Hao and G. W. Hanson, Infrared and optical properties of carbon nanotube dipole antennas, *IEEE Transactions on Nanotechnology*, 5(6): 766–775, 2006.

37. W. R. Davis, J. Wilson, S. Mick, J. Xu, H. Hao, C. Mineo, A. M. Sule, M. Steer, and P. D. Franzon, Demystifying 3D ICs: the pros and cons of going vertical, *IEEE Design and Test of Computers*, 22(6): 498–510, 2005.

38. N. Miura, H. Ishikuro, K. Niitsu, T. Sakurai, and T. Kuroda, A 0.14 pJ/b inductive-coupling transceiver with digitally-controlled precise pulse shaping, *IEEE Journal of Solid-State Circuits*, 43(1): 285–291, 2008.

39. N. Miura, D. Mizoguchi, M. Inoue, K. Niitsu, Y. Nakagawa, M. Tago, M. Fukaishi, T. Sakurai, and T. Kuroda, A 1Tb/s 3W inductive-coupling transceiver for inter-chip clock and data link, In *IEEE International Solid-State Circuits Conference*, 2006, pp. 1676–1685.

40. Y. Yoshida, N. Miura, and T. Kuroda, A 2Gb/s bi-directional inter-chip data transceiver with differential inductors for high density inductive channel array, In *IEEE Asian Solid-State Circuits Conference*, 2007, pp. 127–130.

41. N. Miura, Y. Kohama, Y. Sugimori, H. Ishikuro, T. Sakurai, and T. Kuroda, A high-speed inductive-coupling link with burst transmission, *IEEE Journal of Solid-State Circuits*, 44(3): 947–955, 2009.

42. N. Y. Choi, K.-W. Kwon, and J.-H. Chun, A reliable integrating receiver for inductive coupling chip-to-chip communication, In *Proceedings of ITC-CSCC*, 2010.

43. J. Lee, M. Zhu, K. Choi, J.-H. Ahn, and R. Sharma, 3D network-on-chip with wireless links through inductive coupling, In *2011 International SoC Design Conference*, 2011, pp. 353–356.

44. J. Kim, K. Choi, and G. Loh, Exploiting new interconnect technologies in on-chip communication, *IEEE Journal on Emerging and Selected Topics in Circuits and Systems*, 2(2): 124–136, 2012.

7 Inductive Coupling ThruChip Interface for 3D Integration

Noriyuki Miura and Tadahiro Kuroda

CONTENTS

7.1 INTRODUCTION

ThruChip Interface (TCI) is a low-cost wireless version of through-silicon via (TSV). TCI wirelessly communicates over three-dimensionally (3D) stacked chips through inductive coupling between on-chip coils (Figure 7.1) [1]. The interface coils can be drawn by using existing large-scale integrated (LSI) metal interconnections, and thus TCI is standard complementary metal oxide semiconductor (CMOS) compatible. Unlike TSV, no additional wafer process steps are required, and hence no additional fabrication cost. In addition, the coils are covered under an LSI passivation layer and not exposed for any mechanical contacts to the chip outside. A highly capacitive additional electro-static discharge (ESD) protection circuit can be removed, resulting in small channel loading. As a result, TCI can provide competitive communication performance to TSV even though TCI utilizes a wireless channel (Figure 7.2). TCI is a circuit solution. Fusion combination between the inductive coupling channel characteristics and legacy wireline

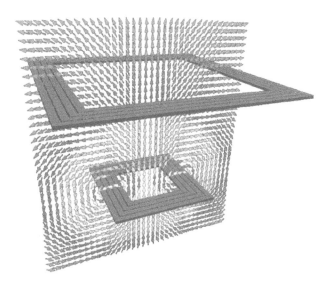

FIGURE 7.1   ThruChip Interface (TCI).

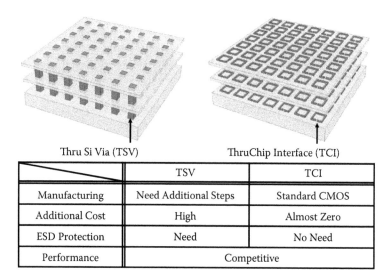

Thru Si Via (TSV)   ThruChip Interface (TCI)

	TSV	TCI
Manufacturing	Need Additional Steps	Standard CMOS
Additional Cost	High	Almost Zero
ESD Protection	Need	No Need
Performance	Competitive	

FIGURE 7.2   TSV vs. TCI.

circuit techniques further enhances TCI communication performance. Utilizing capacitive coupling is one of the alternative solutions to the wireless interface between stacked chips [2]. However, the capacitive coupling has a limitation in using stacked chip communication. It can be applied only to two chips stacked face-to-face. In a field plot of the capacitive coupling shown in Figure 7.3(a), the vertical electric field is shielded by the silicon substrate in the stacked chips. On the other

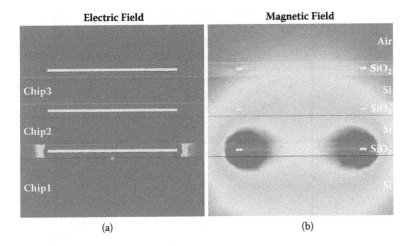

FIGURE 7.3   Electric and magnetic field plot of (a) capacitive and (b) inductive coupling.

hand, the magnetic field of the inductive coupling can penetrate through the silicon substrate (Figure 7.3(b)). Although there is a small eddy current loss in the substrate, the inductive coupling can communicate through the chips. That is the reason why this technology is called ThruChip Interface. In this chapter, the TCI technology is comprehensively overviewed together with recent research activities. In Section 7.2, TCI basics and fundamentals are overviewed. Detailed channel characteristics and design guidelines are explained. In Section 7.3, TCI circuit techniques inspired by legacy wireline arts are introduced for communication performance enhancement. In Section 7.4, TCI applications are described. This section covers from practical to emerging applications. Finally, in Section 7.5, a summary will be given.

7.2   TCI FUNDAMENTALS

7.2.1   CHANNEL DESIGN

A wireless channel utilized in TCI communications is a near-field inductive coupling channel. Wireless channels can be roughly categorized into two groups: one is a near-field and the other is a far-field channel according to signal frequency (i.e., wavelength) and communication distance (Figure 7.4). When the communication distance is longer than the wavelength, an antenna transmits and receives far field and inversely near field for communication. For example, wireless local area network (WLAN) and cell phones utilize a several GHz frequency band (several 10 cm wavelength range) and communicate over the several 10 m ~ km range. Therefore, they are categorized into the far-field channel. On the other hand, TCI utilizes a several 10 GHz frequency band (several centimeters wavelength range) but communicates over a several 10 μm close-proximity communication distance. The TCI channel is therefore categorized in the near-field channel. The biggest advantage of near field is small interference between channels (cross talk). The near field is reactive.

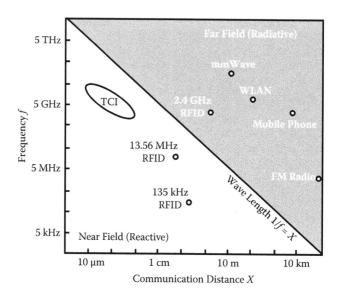

FIGURE 7.4 Wireless technology mapping on frequency and distance plane.

Upon a current flow in a transmitter coil, magnetic flux is instantaneously generated and promptly vanishes upon the current cutoff. On the other hand, the far field is radiative. The magnetic field is generated by the current flow and the field radiates over a long distance. Figure 7.5 shows the magnetic field generated by the coil in a cross-sectional view and a horizontal view. Near field (Figure 7.5(a)) and far field (Figure 7.5(b)) are compared to each other. The magnetic near field is generated concentrically around the coil periphery. The magnetic far field is, on the other hand, radiated far away from the coil. The near-field channel is unsuitable for the long-distance communication; however, the cross talk is small instead. Therefore, the near-field channels can be arranged in high density for high-bandwidth communication [3–5].

The TCI channel operates based on electromagnetic induction. The magnetic field change caused by the transmit current I_T in a transmitter (Tx) coil induces voltage at a receiver (Rx) coil. In an ideal inductive coupling, as shown in Figure 7.6, the received voltage V_R is given by

$$V_R = M \frac{dI_T}{dt},$$ (7.1)

where M is mutual inductance between the Tx and Rx coils. When the transmit current changes in step response, the received voltage V_R becomes a Gaussian pulse (Figure 7.7(a)) and is expressed as

$$V_R(t) = V_P \exp\left[-\frac{4t^2}{\tau^2}\right],$$ (7.2)

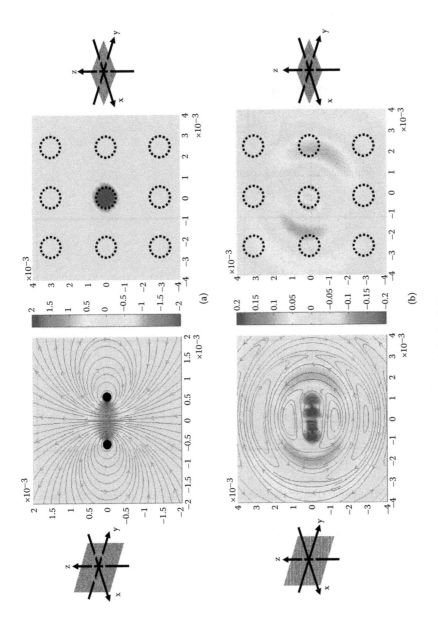

FIGURE 7.5   Magnetic field generated by coil in (a) near field and (b) far field.

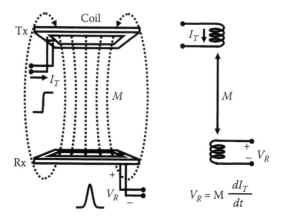

FIGURE 7.6   Ideal inductive-coupling channel model.

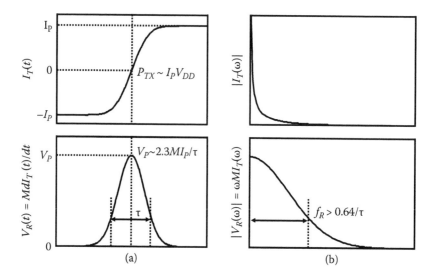

FIGURE 7.7   Transmit and received pulse in (a) time domain and (b) frequency domain.

where τ is a pulse width and V_P is a pulse amplitude. τ limits the maximum data rate below $1/2\tau$ (b/s). V_P is given by

$$V_P = \frac{4}{\sqrt{\pi}} M \frac{I_P}{\tau} \approx 2.3M \frac{I_P}{\tau},\qquad(7.3)$$

where I_P is a transmit current amplitude. Power consumption of the transmitter P_{TX} is approximately calculated by

$$P_{TX} \approx I_P V_{DD} - \alpha\tau\, I_P V_{DD},\qquad(7.4)$$

where V_{DD} is a supply voltage and α is switching activity. Equations (7.3) and (7.4) denote that there is a linear relationship between V_P and P_{TX}. In the frequency domain (Figure 7.7(b)), the V_R frequency spectrum also becomes Gaussian:

$$\left|V_R(\omega)\right| = \frac{\sqrt{\pi}\tau V_P}{2}\exp\left(-\frac{\omega^2\tau^2}{16}\right).$$ (7.5)

The band of interest is up to the frequency f_R, where the power spectrum density drops down to $1/e$ the peak power; then,

$$f_R = \frac{2}{\pi\tau} \approx \frac{0.64}{\tau}.$$ (7.6)

The channel bandwidth should be designed to be higher than f_R; otherwise, the V_R pulse is attenuated or distorted.

Since the coil is made by LSI metal interconnections, an actual inductive coupling channel suffers from bandwidth limitation caused by parasitics of metal interconnections. Figure 7.8 depicts an equivalent circuit model of the actual inductive coupling channel, including parasitic capacitance and resistance of the coil [6]. In this equivalent circuit model, the transimpedance of the channel Z_{IND} is given by

$$Z_{IND} = \frac{V_R}{I_T},$$ (7.7)

$$= \frac{1}{(1-\omega^2 L_R C_R) + j\omega R_R C_R} \cdot j\omega M \cdot \frac{1}{(1-\omega^2 L_T C_T) + j\omega R_T C_T},$$ (7.8)

$$= B_R(\omega) \cdot j\omega M \cdot B_T(\omega),$$ (7.9)

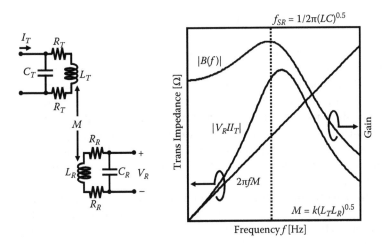

FIGURE 7.8   Equivalent circuit and frequency characteristics of TCI channel.

where L, C, and R are self-inductance, parasitic capacitance, and resistance of the coil. T and R subscripts represent the coil in the transmitter and the receiver, respectively. The second term in Equation (7.8) gives the ideal inductive coupling channel response. It functions as transimpedance with first-order differentiator and exhibits linear characteristics in frequency characteristics (Figure 7.8). The first and third terms in Equation (7.8) denote the bandwidth limitation of the channel $B(\omega)$, which acts as a second-order low-pass filter. $B(\omega)$ has a cutoff frequency and peaking at a self-resonant frequency f_{SR}:

$$f_{SR} = \frac{1}{2\pi\sqrt{LC}}. \tag{7.10}$$

In total, the frequency characteristics of the actual inductive coupling channel become a band-pass filter. The cutoff frequency f_{SR} is designed to be higher than f_R. However, due to the peaking at self-resonance, the frequency characteristics start deviating from the ideal characteristics by approaching f_{SR}. Typically, f_{SR} is designed to be 1.5 ~ 2 times f_R.

The channel gain is determined by the mutual inductance M, which can be rewritten by

$$M = k\sqrt{L_T L_R}, \tag{7.11}$$

where k is a coupling coefficient between the transmitter and the receiver coils. Since L is limited due to bandwidth limitation, k finally governs the channel gain. k is mostly determined by the physical dimensions, ratio between the communication distance X, and coil diameter D and is approximately given as

$$k \approx \left\{ \frac{0.25}{\left(X/D\right)^2 + 0.25} \right\}^{1.5}. \tag{7.12}$$

The calculation results are plotted in Figure 7.9. When X increases $> D$, k is strongly attenuated by X/D^{-3}. The attenuation mitigates below $X/D < 1/2$. Typically, TCI is designed at around $X/D = 1/3$, as the k attenuation is only by X/D^{-1} around this regime. The coil diameter D is set to three times the communication distance $3X$, including margin for chip misalignment and chip thickness variation. The coil turns are maximized as long as f_{SR} drops down to 1.5 ~ 2f_R to maximize the self-inductance $L_T L_R$. The coil line width is adjusted to keep the coil quality factor around 1.5 ~ 2.5. The coil line space is adjusted for fine-tuning of f_{SR}.

For high-bandwidth stacked chip communication, many TCI channels are arranged in parallel. Cross talk between the channels would degrade signal integrity. The channel pitch should be carefully designed for high-density channel arrangement. Figure 7.10 depicts an equivalent circuit model for cross talk analysis in the channel array [7]. Cross talk between the channels is caused by mutual inductance

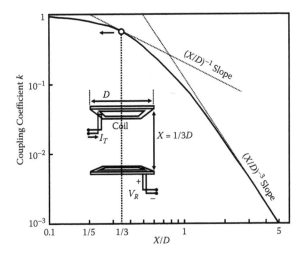

FIGURE 7.9 Calculated coupling coefficient dependence on X/D.

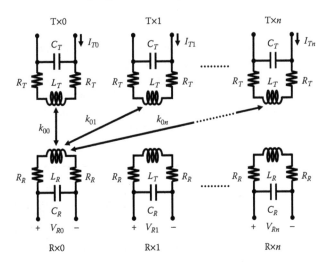

FIGURE 7.10 Equivalent circuit model of inductive-coupling channel array.

between the channels. The received voltage V_{Ri} in the ith Rx coil in an n-channel array is given by

$$V_{Ri} = Z_{IND,k=1} \left(k_{ii} I_{Ti} + \sum_{\substack{j=1 \\ j \neq i}}^{n} k_{ij} I_{Tj} \right). \tag{7.13}$$

where k_{ij} is the coupling coefficient between the ith Rx coil and the jth Tx coil, I_{Tj} is the transmit current in the jth Tx coil, and $Z_{IND,k=1}$ is transimpedance of the channel when $k = 1$. The first term in Equation (7.13) denotes signal, and the second term denotes cross talk components. The cross talk-to-signal ratio (CSR) is thus expressed as

$$\text{CSR} = \left| \frac{\sum\limits_{j \neq i}^{n} k_{ij} I_{Tj}}{k_{ii} I_{Ti}} \right|. \tag{7.14}$$

Since I_T depends on the transmit data transitions, CSR dynamically changes according to the data sequence. Assuming the transmitters in the channel array are all identical, I_T amplitude is constant, and only the polarity changes. Therefore, I_{Tj}/I_{Ti} would take a −1, 0, +1 value. Here, the worst-case cross talk is discussed where all the transmitters transmit the same data transition simultaneously. The worst-case CSR is given as

$$\text{CSR} = \frac{\sum\limits_{j \neq i}^{n} |k_{ij}|}{|k_{ii}|}, \tag{7.15}$$

in the form of a ratio between the coupling coefficient of the main channel and other channels. This means that CSR does not depend on the transmit power because increasing transmit power increases signal and cross talk simultaneously. Each coupling coefficient can be derived from electro-magnetic (EM) simulation. Figure 7.11 presents the simulated results. Mentioned above, since the TCI channel utilizes near-field inductive coupling, the cross talk is rapidly reduced by the cubic of the horizontal

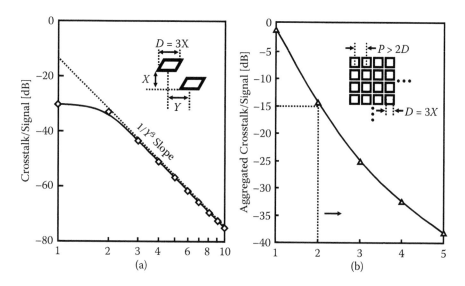

FIGURE 7.11 Simulated cross talk dependence on (a) normalized horizontal distance Y/D and (b) normalized channel pitch P/D.

distance $1/Y^3$ (Figure 7.11(a)). The worst-case aggregated cross talk in the channel array is also rapidly reduced by increasing the channel pitch P (Figure 7.11(b)). The minimum channel pitch is twice the coil diameter in the typical design. The coil diameter is three times the communication distance. The minimum channel pitch is therefore six times the communication distance. For high-density and high-bandwidth communication, a communication distance reduction is required. A 3D scaling scenario, including chip thickness, is the key for performance improvement of TCI [8].

7.2.2 TRANSCEIVER CIRCUIT DESIGN

The TCI transceiver circuit can be implemented in a very simple, almost CMOS digital circuit (Figure 7.12) [9]. The transmitter consists of two inverters to form an H-bridge driver circuit. It converts differential transmit digital data *Txdata* into the transmit current I_T and directly drives the transmitter coil. When *Txdata* is low, NMOS in the right inverter and PMOS in the left inverter are ON and positive transmit current $-I_p$ flows in the transmitter coil. When *Txdata* toggles to high, NMOS in left and PMOS in right are ON and the transmit current also reverses to $+I_p$. Through the inductive coupling channel, the received voltage V_R is proportional to dI_T/d_t. Thus, positive and negative pulse-shaped voltage is induced in the receiver coil at every data transition. Positive pulse is generated when *Txdata* transits from low to high, and vice versa. The receiver is a hysteresis comparator. It recovers digital data *Rxdata* from the pulse-shaped voltage V_R. The hysteresis comparator consists of two stages: one is the gain stage for data transition and the other is the latch stage for data hold. The second latch stage (cross-coupled MOS pair) shifts the threshold voltage of the comparator either positive or negative according to the holding data *Rxdata*. When *Rxdata* is low, the threshold voltage (dashed line in Figure 7.12) shifts slightly positive. Upon the positive received pulse input, the first gain stage forces a flip in

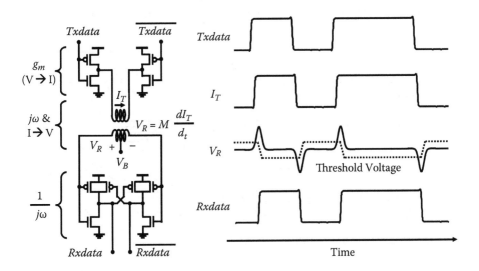

FIGURE 7.12   TCI transceiver circuit and its operation waveform.

the holding data in the latch stage to high. The latch stage then shifts the threshold voltage negative and holds data high until the negative pulse input. By repeating this operation, digital data can be delivered through the inductive coupling channel. To see this communication system in terms of circuit functionality, the transmitter behaves as transconductance g_m (voltage-to-current converter), the inductive coupling channel functions as transimpedance (current-to-voltage conversion) with first-order differentiation ($j\omega$), and the receiver operates as a first-order integrator ($1/j\omega$) by the positive feedback in the cross-coupled NMOS latch stage. All through this system, channel response becomes ideally flat from *Txdata* to *Rxdata*.

7.3 TCI CIRCUIT TECHNIQUES

From the telephone exchange board era to the optical fiber network of today, various circuit techniques were innovated in a wireline circuit field. Fusion combination between TCI and wireline technology introduces additional functions and also enhances communication performance. In this section, several TCI circuit techniques inspired by legacy wireline arts are introduced by discussing some specific examples.

7.3.1 TCI Repeater

In wireline, a repeater circuit technique is used for long-distance communication (Figure 7.13(a)). When the data are delivered from point A to B over a long wire, the data signal is gradually attenuated due to parasitic resistance and capacitance of the wire, and finally, the communication becomes difficult. To solve this problem, a repeater (e.g., CMOS inverter buffer) is inserted in the long wire and relays the signal. The repeater amplifies and recovers the signal amplitude to original full swing and enables long-distance communication. A similar technique can be applied to TCI (Figure 7.13(b)). For communication between chips stacked far apart from each other in multiple-chip stacking, a TCI repeater is integrated on the chip in between relays

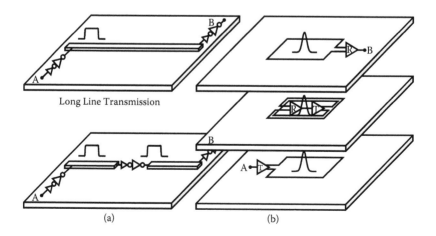

FIGURE 7.13   (a) Wireline repeater and (b) TCI repeater.

of the data signal in a manner similar to that of the wireline. The biggest difference to the wireline repeater is that in the TCI repeater, the transmitter transmits the signal both upward and downward simultaneously (the TCI transmitter cannot define the signal direction). In the case of Figure 7.13(b), TCI in the bottom chip transmits the data and the TCI repeater in the middle relays the data to the top chip, but at the same time, the repeater returns the signal down to the bottom chip (like echo). This vertical echo cross talk can be ignored by properly delaying the timing of transmitting the next pulse signal, as in Figure 7.14. Since the polarity of the echo cross talk (C_{22} and C_{32} in Figure 7.14) is identical to that of the actual signal, the holding data in the hysteresis receiver are not distorted. Unless the next signal pulse collides with the echo cross talk, the TCI repeater can correctly relay the data for long-distance communication [10].

However, because of delaying the transmission timing of the next pulse, the data rate will be reduced to less than one-third of the maximum data rate of TCI. This will be allowed in case the data rate is restricted by other peripheral circuits. In fact, often the data rate is limited by a multiplexer (MUX) and a demultiplexer (DMUX) implemented in CMOS for energy efficiency. In a 65 nm CMOS, the TCI potential data rate is more than 30 Gb/s [11]. MUX/DMUX in 65 nm CMOS implementation is around 8 Gb/s. If a current-mode logic (CML) is employed in MUX/DMUX circuits, it is possible to exploit the maximum available data rate of TCI. A three coils/link TCI repeater [12] can be utilized for high-speed relayed transmission (Figure 7.15). In each chip, three pairs of TCI transceivers are implemented. The repeater link is formed by electrically switching ON and OFF the transmitter and the receiver according to the stacked chip address. The echo cross talk comes from the transmitter at a distance twice that for the actual signal link, and therefore becomes small enough to be ignored by the receiver hysteresis.

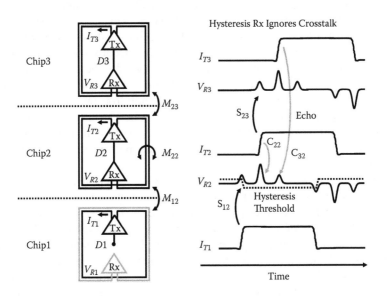

FIGURE 7.14 TCI repeater circuit.

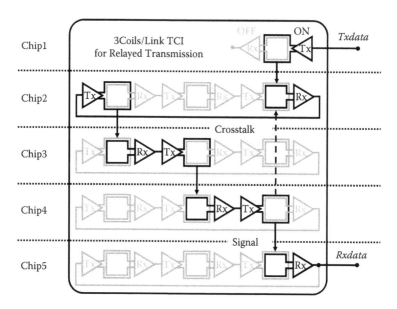

FIGURE 7.15   3 coils/link TCI repeater.

7.3.2   TCI MULTIDROP BUS

As shown in Figure 7.16(a), a multidrop bus structure is utilized to broadcast the same data to multiple drop points. It is also possible to realize the multidrop bus with TCI as shown in Figure 7.16(b). A magnetic field can penetrate through multiple chips. By properly increasing the coil diameter according to the maximum communication distance, multiple coils can be inductively coupled to form the multidrop bus structure. In the wireline multidrop bus with either wire bonding or TSV, ESD protection circuits significantly increase the channel loading as the number of drop points (in other words, the number of stacked chips, N_{STACK}) increases, resulting in significant power and delay degradation (Figure 7.17). On the other hand, in the TCI multidrop bus, the channel load at each drop point is originally small, and it can be partially seen through the inductive coupling channel. Therefore, the power and delay penalty is small even if the number of stacked chips increases. TCI is suitable for forming the multidrop bus structure.

Combination of this TCI multidrop bus and the TCI repeater mentioned in the previous section introduces a trade-off between power and area of TCI. As shown in Figure 7.18, when multiple chips are stacked, relayed transmission at each chip increases the number of the TCI repeaters, and hence power dissipation. By increasing the coil diameter, for example, by eight times, forming a multidrop bus among eight chips, and relaying at every eighth chip, the number of repeaters can be reduced to one-eighth and power dissipation to one-eighth. A technical challenge is in how to lay out the large-diameter coil for the multidrop bus. For a logic chip, a coil layout using two different metal layers for horizontal (X) and vertical (Y) coil wires,

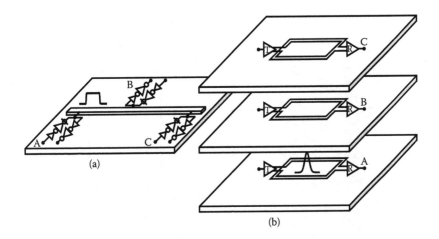

FIGURE 7.16 (a) Wireline multidrop bus and (b) TCI multidrop bus.

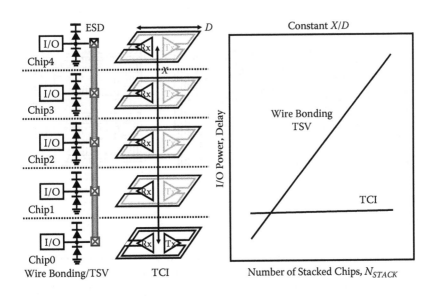

FIGURE 7.17 I/O power and delay dependence on N_{STACK} in multidrop bus.

namely, XY coil layout style [13], can be utilized (Figure 7.19). For example, by using the fourth (even) metal layer for horizontal coil wire and the fifth (odd) metal layer for vertical coil wire, other logic interconnections can go through under or above the coil. The coil can be hidden in the sea of logic interconnections to save the layout area penalty in arranging the large coil on the chip. For a memory chip, the large coil can be placed over the memory core, which typically occupies most of the chip area (Figure 7.20). By partially removing the third metal for power reinforcement, the large coil can be placed with almost no area penalty on the memory chip [14].

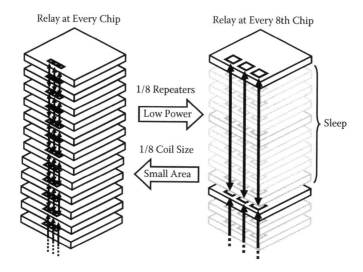

FIGURE 7.18 Trade-off between power and area using TCI repeater and multidrop bus.

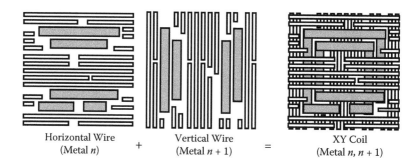

FIGURE 7.19 XY coil layout style.

One problem is magnetic field loss due to eddy current in densely drawn bit/word lines on the memory core. The magnetic field still can penetrate the bit/word lines; however, additional transmit power is needed for compensating the eddy current loss. Increasing the number of chips penetration reduces the number of repeaters, while increasing loss and actually increasing additional transmit power at each repeater. There is an optimal number of chips penetration to minimize total power consumption of the TCI multidrop bus. The optimal number of chips penetration depends on the layout density of the memory bit/word lines. Figure 7.21 presents measured data using a commercial 43 nm NAND Flash memory core. The memory cores are sandwiched by the TCI transmitter and the receiver chip. By changing the number of the stacked memory core, the relationship between the additional transmit power and the number of chips penetration is measured. In this particular case, using a 43 nm NAND Flash, the optimal number of chips penetration was measured to be eight [14].

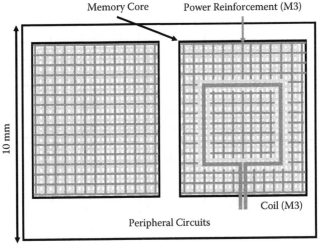

FIGURE 7.20 TCI coil over memory core.

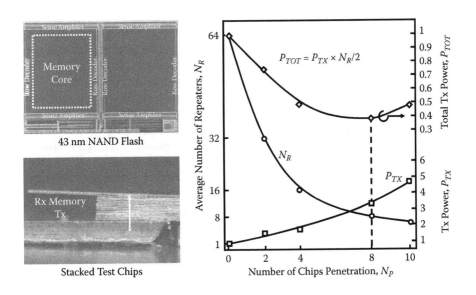

FIGURE 7.21 Measured additional transmit power vs. number of chips penetration.

7.3.3 SOURCE-SYNCHRONOUS BURST TRANSMISSION

Since channel load of TCI is equivalent to <1 mm long on-chip interconnection, TCI has a potential advantage in high-speed operation compared to a conventional wireline interface that suffers from heavy off-chip channel loading, including ESD protection circuits. Therefore, as shown in Figure 7.22, conventional parallel

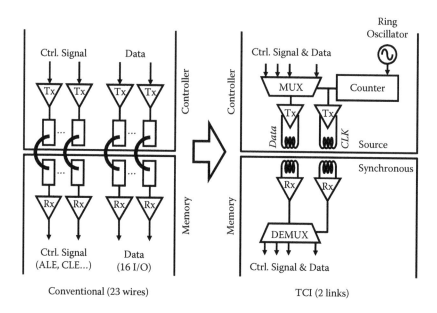

FIGURE 7.22   TCI burst transmission for compact I/O.

interfaces can be multiplexed into one TCI for a compact I/O interface. In addition, by adding header bits and employing packet communication, not only data but also command and address can be transferred by the same single TCI. Even if the single TCI is larger than the conventional interface, the data rate per unit area (i.e., area efficiency) can be improved by multiplexing. However, in order to multiplex low-speed data into a high-speed data stream, a high-frequency clock is required. One way is to use phase-locked loop (PLL) to generate the high-frequency clock, but large area and power overhead are problems. Burst transmission with high-frequency burst clock generation by a local ring oscillator is one of the solutions for area-efficient high-speed data multiplexing [15]. The high-frequency clock used for multiplexing data is also transferred by using the same TCI transceiver in parallel with the data transceiver. This source-synchronous transmission significantly reduces timing skew between the data and the clock, enabling robust high-speed burst transmission.

7.3.4   INJECTION LOCKING CDR

Clock and data recovery (CDR) is one of the most significant wireline circuit innovations. A synchronous clock used for retiming the received data is recovered from the received data itself. CDR is utilized for more than 10 Gb/s high-speed communications. Among them, injection locking CDR is area and power efficient and often used for high-speed burst transmission [16]. The injection locking CDR employs an edge detector to extract the clock frequency information from received data. In Figure 7.23(a), the extracted edge pulse is injected into a local oscillator to recover the clock synchronized with the received data. Injection locking CDR only requires the edge detector and the local oscillator with several injection transistors.

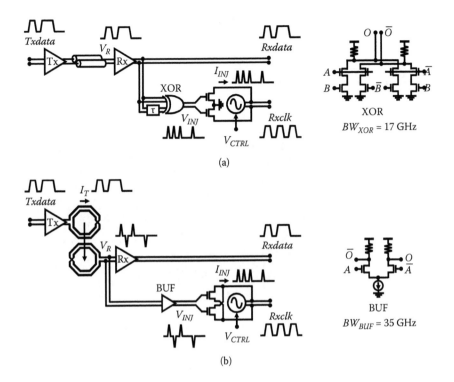

FIGURE 7.23 Injection locking CDR in (a) wireline and (b) TCI.

Therefore, small power and area CDR can be realized; however, the operation speed is limited in conventional injection locking CDR where the edge detector consists of XOR and delay buffers. Due to the stacked transistor topology in the XOR circuit, the bandwidth BW_{XOR} is only 17 GHz in a 65 nm CMOS. As a result, the possible data rate is limited to 17 Gb/s. Although the receiver clock frequency is half-rate, the edge detector operates at full-rate clock frequency and thus becomes the critical path in this CDR. By utilizing the derivative property of inductive coupling, the edge detection can be performed by the inductive coupling channel instead of XOR for high-speed operation. Since the coil is just like on-chip interconnection, the potential bandwidth of the inductive coupling channel can be much higher than 100 GHz, and the channel does not restrict the data rate. The injection circuits consist of two parallel NMOS transistors connected to the VCO outputs. Driven by the differential edge signal V_{INJ}, one of the NMOS transistors is switched ON, and it shunts the VCO outputs to perform injection locking. In an actual implementation, an isolation buffer BUF is inserted between the receiver coil and the injection circuits in order to avoid clock feedthrough to the data receiver. This buffer insertion does not degrade the operation speed of the CDR because the bandwidth of the buffer BW_{BUF} is higher than that of the data receiver (hysteresis comparator) and is as high as 35 GHz in a 65 nm CMOS. A test chip is designed and fabricated in a 65 nm CMOS [11]. Successful operation up to 30 Gb/s is confirmed in this test chip measurement (Figure 7.24).

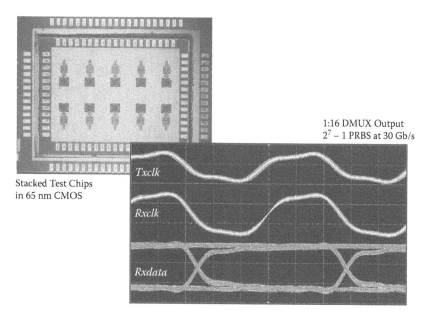

1:16 DMUX Output
$2^7 - 1$ PRBS at 30 Gb/s

Stacked Test Chips
in 65 nm CMOS

FIGURE 7.24   Test chip in 65 nm CMOS and waveform snapshot of 30Gb/s CDR.

7.3.5 COUPLED-RESONANT CLOCK DISTRIBUTION

Coupled resonation is one of the well-known physical phenomena. The frequency and phase of the multiple oscillators are synchronized through weak coupling between them. It can also be seen in nature, like synchronous flashing of fireflies, and has been ever deeply analyzed in the circuit field [17]. One of the applications of this physical phenomenon is on-chip clock distribution [18]. Multiple ring oscillators are output tied (coupled) together to lock the clock frequency and phase among all the ring oscillators by coupled resonation as illustrated in Figure 7.25(a). By tapping out the clock from the oscillators-coupling wire grid, a small-skew clock signal can be distributed across the entire chip area. Compared to a conventional H-tree-based clock distribution, this coupled-resonant clock distribution can improve variation immunity and clock distribution skew at higher clock frequency. Small-skew clock distribution is of course important in 3D ICs, but very difficult. Even by using TSV the clock delay between the stacked chips is not ignorable. Even if distributed PLLs are utilized among the 3D stacked chips, the reference clock skew is visible, and hence larger clock skew appears between the stacked chips.

The TCI inductive coupling link can be utilized for the clock distribution among 3D stacked chips. Ideally, as Figure 7.25(b) shows, by integrating an LC oscillator in each stacked chip and coupling coils in the oscillator, a no-skew clock can be distributed among the 3D stacked chips. By using this coupled LC oscillator for the vertical reference clock distribution and the coupled ring oscillators for the horizontal clock distribution, very-low-skew 3D clock distribution can be realized. Test chip measurement in a 0.18 μm CMOS is reported in [19] (Figure 7.26). Less than 4% unit interval (UI) at 1 GHz 3D clock distribution has been confirmed (Figure 7.27).

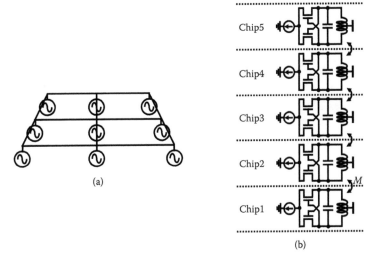

FIGURE 7.25 Coupled-resonant clock distribution using (a) ring oscillators and (b) LC oscillators.

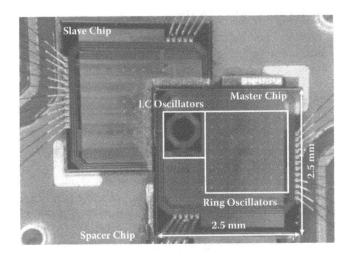

FIGURE 7.26 Stacked test chips for 3D clock distribution.

7.4 TCI APPLICATIONS

TCI applications are widely spread across various fields. For example, 3D stacking of processors and memories [20–22], noncontact wafer testing [23], and an on-chip bus probe monitoring system [24] are reported. In this section, as representative TCI applications, NAND Flash memory stacking from commercially practical application fields and noncontact, and hence long-lifetime, memory from future emerging applications are introduced.

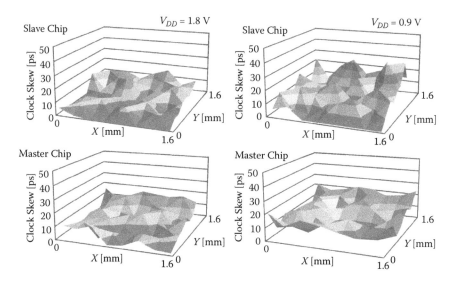

FIGURE 7.27   Measured clock skew in 3D clock distribution.

7.4.1   NAND FLASH MEMORY STACKING

NAND Flash memory stacking is one of the most advanced techniques in current 3D stacking, and hence one of the killer applications of TCI. Typical solid-state drive (SSD) mounts 8 ~ 16 NAND Flash memory packages on the board, and each package houses around eight NAND Flash memory chip stacks. In total, as many as 128 NAND Flash memory chips are utilized in a single SSD. In order to integrate all these 128 NAND Flash memory chips in a single package, 30 bonding wires per chip, and thus almost 4,000 wires in total, have to be routed in one package. This is very difficult. These bonding wires are used for delivering data signals and supplying power inside and outside of the package. Among them, bonding wires for the data signals inside the package can be replaced by the TCI repeater and the multidrop bus (Figure 7.28). The total number of the required bonding wires can be reduced to around 500, which is 1/8 of the conventional. This makes it possible to integrate 128 NAND Flash memory chips in a single package. It is reported in [10, 14] that 1 TByte memory capacity SSD can be realized with 1/80 assembly volume and 1/15 I/O power dissipation compared to conventional wired SSD.

7.4.2   NONCONTACT MEMORY CARD AND PERMANENT MEMORY

By not only delivering data but also supplying power through the inductive coupling channel, completely noncontact semiconductor devices can be realized. One of the application examples is a noncontact memory card (Figure 7.29). In a conventional memory card, metal connectors for data communication and power supply are exposed. Through longtime repetition of attach and remove with the host device, the connectors are worn off, making it difficult to access the internal memory chip. A completely non-contact memory by using TCI removes all the mechanical connectors and can guarantee

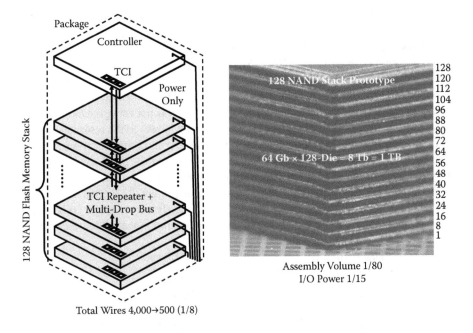

FIGURE 7.28   128 NAND Flash memory stacking with TCI.

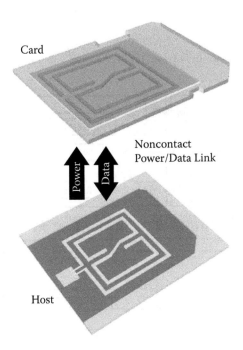

FIGURE 7.29   Noncontact memory card.

longtime reliability of the memory card. In addition, similar to the 3D stacked chips, since this is wireless, ESD protection devices can be removed, resulting in high-speed card access. A technical challenge lies in interference from the power channel to the communication channel (Figure 7.30). Since the input power between them is three orders of magnitude different (60 dB larger input power in the power channel), more than 80 dB isolation between them is needed. Although there is 40 dB difference in the operating frequency, filtering out in a frequency domain is not enough. One way is to interleave in a time domain (Figure 7.31). By delivering power to the reservoir capacitor in the memory card and stopping delivering power during communication, interference between the power and the communication channels can be eliminated. However, the effective data rate is reduced due to a limited duty ratio of data communication. A differential coil structure can solve this problem (Figure 7.32) [25]. This structure is a kind of space division multiplexing technique. Two rectangular coils turned inversely are connected together to introduce directionality in the inductive coupling.

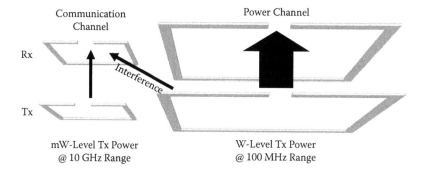

FIGURE 7.30   Interference from power link to data link.

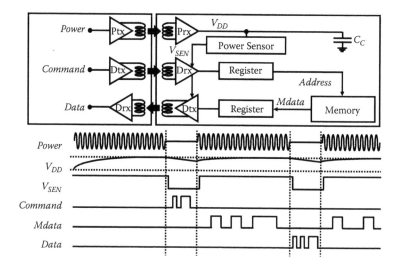

FIGURE 7.31   Time interleaving scheme.

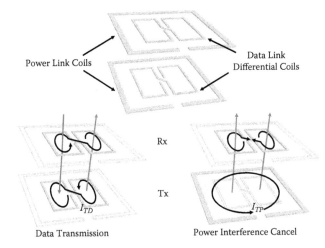

FIGURE 7.32 Differential coil scheme.

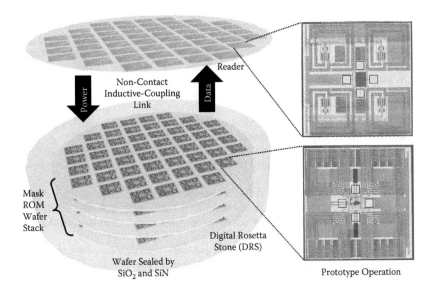

FIGURE 7.33 Permanent memory system: Digital Rosetta Stone (DRS).

By nesting the differential coil in the center of the power coil, the interference from the power coil can be eliminated since the magnetic field generated by the power coil induces the same voltage in each the right and the left subcoil. The subcoils are inversely turned and tied together. The interference from the power channel can be finally canceled out at the differential coil terminals.

By employing a similar noncontact scheme, a completely shielded permanent memory, namely, Digital Rosetta Stone (DRS), can be realized (Figure 7.33).

Memory data are stored in mask read-only memories (ROMs) on entire silicon wafers by electronic beam scanning. Since the power and data can be delivered wirelessly, this mask ROM wafer can be completely shielded by stable protection film such as SiO_2 or SiN. By avoiding erosion of oxygen and water into the wafer, more than 1,000 years of lifetime permanent memory can be realized. Theoretical analysis and prototype demonstration are reported in [26].

7.5 SUMMARY

TCI is a low-cost wireless version of TSV. TCI utilizes a near-field inductive coupling channel to provide high-density, high-speed, and low-power 3D stacked chip communication. Fusion combination between the inductive coupling channel and legacy wireline circuit techniques further enhances communication performance. TCI is ready for practical use. TCI can be applied to practical applications such as NAND Flash memory stacking and can possibly open up emerging applications such as a noncontact memory card and a permanent memory system.

REFERENCES

1. D. Mizoguchi, et al., A 1.2 Gb/s/pin wireless superconnect based on inductive inter-chip signaling (IIS), *ISSCC Digest of Technical Papers*, February 2004, pp. 142–143.
2. K. Kanda, et al., 1.27 Gb/s/pin 3 mW/pin wireless superconnect (WSC) interface scheme, *ISSCC Digest of Technical Papers*, February 2003, pp. 186–187.
3. N. Miura, et al., A 195 Gb/s 1.2 W 3D-stacked inductive inter-chip wireless superconnect with transmit power control scheme, *ISSCC Digest of Technical Papers*, February 2005, pp. 264–265.
4. N. Miura, et al., A 1Tb/s 3 W inductive-coupling transceiver for inter-chip clock and data link, *ISSCC Digest of Technical Papers*, February 2006, pp. 424–425.
5. N. Miura, et al., An 8 Tb/s 1pJ/b 0.8 mm²/Tb/s QDR inductive-coupling interface between 65 nm CMOS and 0.1μm DRAM, *ISSCC Digest of Technical Papers*, February 2010, pp. 436–437.
6. N. Miura, et al., Analysis and design of inductive coupling and transceiver circuit for inductive inter-chip wireless superconnect, *IEEE JSSC*, 40(4): 829–837, 2005.
7. N. Miura, et al., Crosstalk countermeasures for high-density inductive-coupling channel array, *IEEE JSSC*, 42(2): 410–421, 2007.
8. T. Kuroda, et al., Perspective of low-power and high-speed wireless inter-chip communications for SiP integration, *ESSCC Digest of Technical Papers*, September 2006, pp. 3–6.
9. N. Miura, et al., An 11 Gb/s inductive-coupling link with burst transmission, *ISSCC Digest of Technical Papers*, February 2008, pp. 298–299.
10. N. Miura, et al., A 2.7 Gb/s/mm² 0.9 pJ/b/chip 1 coil/channel ThruChip interface with coupled-resonator-based CDR for NAND flash memory stacking, *ISSCC Digest of Technical Papers*, February 2011, pp. 490–491.
11. Y. Take, et al., A 30 Gb/s/link 2.2 Tb/s/mm² inductively-coupled injection-locking CDR, *A-SSCC Digest of Technical Papers*, November 2010, pp. 81–84.
12. M. Saito, et al., 47% Power reduction and 91% area reduction in inductive-coupling programmable bus for NAND flash memory stacking, *Transactions on Circuits and Systems I*, 57(9): 2269–2278, 2010.
13. M. Saito, et al., An extended XY coil for noise reduction in inductive-coupling link, *A-SSCC Digest of Technical Papers*, November 2009, pp. 305–308.

14. M. Saito, et al., A 2 Gb/s 1.8 pJ/b/chip inductive-coupling through-chip bus for 128-Die NAND-flash memory stacking, *ISSCC Digest of Technical Papers*, February 2010, pp. 440–441.

15. N. Miura, et al., A high-speed inductive-coupling link with burst transmission, *IEEE JSSC*, 44(3): 947–955, 2009.

16. M. Nogawa, et al., A 10 Gb/s burst-mode CDR IC in 0.13 μm CMOS, *ISSCC Digest of Technical Papers*, February 2005, pp. 228–229.

17. R. Adler, A study of locking phenomena in oscillators, *Proceedings of IEEE*, 60: 1380–1385, 1973.

18. M. Mizuno, et al., A noise-immune GHz-clock distribution scheme using synchronous distributed oscillators, *ISSCC Digest of Technical Papers*, February 1998, pp. 404–405.

19. Y. Take, et al., 3D clock distribution using vertically/horizontally coupled resonators, *ISSCC Digest of Technical Papers*, February 2013, pp. 258–259.

20. K. Niitsu, et al., An inductive-coupling link for 3D integration of a 90 nm CMOS processor and a 65 nm CMOS SRAM, *ISSCC Digest of Technical Papers*, February 2009, pp. 480–481.

21. Y. Kohama, et al., A scalable 3D processor by homogeneous chip stacking with inductive-coupling link, *Symposium on VLSI Circuits Digest of Technical Papers*, June 2009, pp. 94–95.

22. Y. Sugimori, et al., A 2 Gb/s 15 pJ/b/chip inductive-coupling programmable bus for NAND flash memory stacking, *ISSCC Digest of Technical Papers*, February 2009, pp. 244–245.

23. A. Radecki, et al., 6W/25 mm^2 inductive power transfer for non-contact wafer-level testing, *ISSCC Digest of Technical Papers*, February 2011, pp. 230–231.

24. H. Ishikuro, et al., An attachable wireless chip-access interface for arbitrary data rate using pulse-based inductive-coupling through LSI package, *ISSCC Digest of Technical Papers*, February 2007, pp. 360–361.

25. Y. Yuan, et al., Simultaneous 6 Gb/s data and 10 mW power transmission using nested clover coils for non-contact memory card, *Symposium on VLSI Circuits Digest of Technical Papers*, June 2010, pp. 199–200.

26. Y. Yuan, et al., Digital Rosetta Stone: A sealed permanent memory with inductive-coupling power and data link, *Symposium on VLSI Circuits Digest of Technical Papers*, June 2009, pp. 26–27.

8 Fabrication and Modeling of Copper and Carbon Nanotube-Based Through-Silicon Via

Brajesh Kumar Kaushik, Manoj Kumar Majumder, and Archana Kumari

CONTENTS

8.1 INTRODUCTION

During the recent past, the semiconductor industry has faced new challenges in designing of devices and circuits with improved speed and power. Designing and fabrication of novel devices is essential for providing more functionality at higher speed and in smaller dimensions. A 3D integrated circuit (IC) offers an alternative solution to 2D planar ICs by either increasing the device functionality or combining different technologies [1, 2]. For highly complex and large designs and their manufacturing, the 3D integrated IC delivers significant benefits in improving the area, circuit performance, integration density, interconnect power consumption, and heterogeneous technology integration capabilities [3].

8.1.1 3D IC STACKING

In the 1980s, the theoretical studies [4, 5] suggested that the signal delay and power consumption would considerably reduce using a 3D IC. A 3D IC can be considered a chip having multiple tiers of stacked, bonded, and electrically connected thinned-active 2D ICs. The 2D ICs are primarily stacked with vertical vias and filled with silicon or oxide layers. The word *tier* is used to distinguish the transferred layers of a 3D IC from the design and physical layers. It primarily defines the functional unit of a wafer or chip that consists of an active interconnect, silicon, silicon-on-oxide (SOI) wafer, and buried oxide (BOX). From the view of industrial applications, 3D IC stacking technology primarily offers complete coverage of designing, testing, and fabrication processing of through-silicon via (TSV) integrations. The stacking technology is also useful for future industrial applications and cutting-edge design methodologies. Therefore, stacking technology emerges as an essential resource for semiconductor engineers and portable device designers. The stacking technology primarily encapsulates the following research areas:

- High-density through-silicon stacking (TSS) technology
- Technology computer-aided design (TCAD) tool solutions and design automation for TSV-based 3D IC stack
- Assembly and test aspects of TSV technology

- Chemical mechanical polishing (CMP) for TSV applications
- Process integration for TSV manufacturing
- Copper electrodeposition for TSV
- Etching of TSV with high-aspect-ratio silicon substrate
- Practical design ecosystem for heterogeneous 3D IC products
- Dielectric deposition for TSV
- Barrier layer and seed layer deposition
- Temporary and permanent bonding

8.1.2 Basics of TSVs

Different technologies such as wire bonding, flip-chip bonding, and TSVs for the electrical interconnection of stacked dies already exist. Particularly, TSVs can provide a shorter signal path that exhibits superior electrical characteristics in terms of reduced resistive, inductive, and capacitive components. Therefore, the primary focus of current research is toward the designing of reliable and cost-efficient TSVs. In the current nanoscale regime, development of TSV-based technology that connects the 3D stacked ICs has attracted many researchers globally. TSV appears to be one of the greatest technological challenges developed by 3D integration. Beyond TSV considerations, all the conventional design procedures and layout capabilities are unsuitable for 3D integration. Using conventional designing, a few labs or companies had developed 3D-compatible design methodologies [6–8], but they were not commercially available. The well-known methodologies involved in realizing the 3D integration include the face-to-face integration and the back-to-back integration.

Traditionally, the connections were made through the multiple IP cores on a single die (system-on-chip), multiple dies in a single package (multichip package), and multiple ICs on a printed circuit board (PCB). Later on, system-in-package (SiP) technology was introduced where dies containing ICs were stacked vertically on a substrate. Another stacking technique is package-on-package (PoP); it uses vertically stacked multiple packaged chips. The latest development in this area is the 3D stacked IC that aligns at the backside of a thinned-down die. This type of stacking is made by using TSV that employs a single package containing a vertical stack of naked dies and allows the die to be vertically interconnected with another die.

TSVs provide highly integrated 3D packaging using the vertical stacking of chips. It can be referred to as a vertical electrical connection VIA (vertical interconnect access) that passes completely through a silicon wafer or die. TSVs are low-capacity and high-density interconnects in comparison to the traditional wire bonds. They allow more interconnect lines between the vertically stacked dies, and hence they exhibit higher speed and lower power dissipation. Thus, a new generation of "super chips" is introduced using TSV-based vertical interconnects.

8.1.3 Properties of TSVs

TSVs are electrical interconnects that vertically penetrate through stacked wafers or chips. This methodology also effectively reduces the mechanical strain. This section primarily provides the details of electrical and thermomechanical properties.

8.1.3.1 Electrical

For global interconnects, TSVs provide lesser parasitic values in comparison to the typical 2D wires. Due to these reduced parasitics, TSVs are electrically fast and attract system designers. Thus, using 3D TSV-based technologies, the signaling performance and driving capability of integrated circuits would improve significantly.

8.1.3.2 Thermomechanical

A TSV-based 3D technology can deliver reduced mechanical strain in comparison to the traditional 2D wire bond. In 3D interconnect structures, serious reliability issues such as Si cracking and performance degradation of devices can be raised due to the process-induced thermal stresses. The thermal stresses in silicon are reduced as a function of distance from an isolated TSV and increased with the TSV diameter. The thermal expansion mismatch between the constituent materials may lead to the degradation of performance of stress-sensitive devices during the fabrication of TSVs [9]. Additionally, the thermal stresses can drive cracks in silicon substrates to cause device failure.

8.1.4 TSV Applications

In current researched area of interconnects, TSV-based 3D technologies demonstrate a variety of applications in the area of microelectronics and nanoelectronics:

- The application of TSV is found for both the homogeneous and heterogeneous integrations. Homogeneous integration is defined as the connection between the dies of the same type, whereas heterogeneous integration refers to the connection between the dies of different types. Using the homogeneous TSV integration, the cost of the PCB is significantly reduced. It is due to the fact that system integration for a homogenous TSV is used to produce stacked dynamic random access memory (DRAM) that boosts the capacity of memory per unit board area. It results in reduced latency and increased bandwidth between the system processor and memory.

  A TSV-based heterogeneous integration includes its applications as image sensors or communication chips in cell phones that are stacked with memory and digital signal processing (DSP). This scheme is also applicable for 3D gaming systems. The main advantage of using heterogeneous integration is the faster exchange of data between the subcomponents in comparison to the other 3D integration techniques, such as SiP, PoP, etc.

- The image sensor is another important application of TSV. The 3D image sensor with a TSV silicon interposer provides more flexibility and a higher degree of vision-based system integration. The integration does not require any specially designed circuits and components. Therefore, TSV-based 3D integration can be used to integrate all the image sensors and processors available in the market.

8.1.5 Challenges to TSV Implementation

Some fabrication-related issues associated with TSVs may limit their applications. These issues are primarily associated with the manufacturing cost and designing.

8.1.5.1 Cost

The biggest barrier associated with TSV technology is its implementation cost. The cost is primarily determined by numerous aspects of designing and manufacturing. Particularly, the major cost barriers are the bonding/de-bonding and via barrier/filling of TSV.

A TSV exhibits its tremendous value if its production cost perfectly fits with the industrial roadmap. For instance, a TSV-based 3D IC can provide improved electrical characteristics to address the limits in leakage and electrical performance of complementary metal oxide semiconductor (CMOS) ICs beyond 11–16 nm. But these benefits come at a very high cost that can be regarded as a major challenge.

8.1.5.2 Design

In the current applications of 3D vertical interconnections, TSVs can be adopted to combine different types of chips, ICs, and design guidelines to address a variety of issues. The layout and chip architecture have faced fundamental design changes, such as stacking, wire bonding, etc., due to the several thousand interconnections between the dies. Designers for each chip type used in the integration scheme will have to leverage the same master layout to line up connection points between the chips. At the same time, possible heat generation issues also need to be considered. Overheat in the stacked chips may be occurring due to some designing problem related to thermal management mechanisms. Thus, the hot spots and temperature gradients strongly affect the reliability of the chip. Apart from this, longer TSV substantially increases the thermal and intrinsic stress that becomes a major concern in mechanically stable design. The thermomechanical stress arises from the difference between the coefficient of thermal expansion (CTE) of silicon and the interconnection metal, whereas the intrinsic stress results from different physical mechanisms that take place during the metal deposition. Therefore, more sophisticated solutions are required to address the designing issues related to thermal and stress management of a 3D IC.

8.1.5.3 Manufacturing

The cost-effectiveness of designing a TSV follows the full manufacturing sequence that requires integration and optimization between traditional steps in back-end packaging and wafer processing. The entire flow can be optimized to deliver the greatest performance (yield, reliability) for the highest productivity (cost).

For TSV applications, via-first or via-last approaches primarily alter the sequence of manufacturing processes. In either approach, the TSV stack undergoes thinning, bonding, wafer processing on bonded/thinned wafers, and subsequent de-bonding. Wafers are mainly bonded to the glass or dummy silicon and thinned down to a thickness ranging from 30 to 125 μm.

8.1.6   TSV Fabrication Technology

TSVs can be made at different stages of wafer processing, front-end-of-line (FEOL), back-end-of-line (BEOL), or regular fabrication processes. This section provides a detailed description related to different steps of TSV fabrication.

8.1.6.1   Etching

Before filling the TSV material, it is required to etch the trenches deep into the silicon that exhibits an extremely high aspect ratio. The approximate height and width of a trench are 100–150 microns and 1–5 μm, respectively. The etching process of TSV primarily follows a number of methods, such as wet etching, dry etching, laser drilling, reactive ion etching (RIE), deep reactive ion etching (DRIE), etc. [10]. Anisotropic wet etching is generally used for TSV formation with very large pitch. However, isotropic wet etching can be combined with dry etching to adjust the desired profile. Dry etching primarily includes laser drilling and RIE. Laser drilling provides significant cost advantages over patterning and etching by eliminating the lithography steps.

The RIE process usually involves a high-density plasma (HDP) source where the ions and radial species from plasma etch the surface chemically and physically. After etching, the remaining by-products are removed and the plasma becomes continuously reactive. Due to the slow nature of the RIE etching process, scientist Bosch patented a new alternative etching process known as DRIE or the Bosch process. This process follows an alternative etching (using SF_6) and sidewall passivation step (using C_4F_8) with a high etch rate up to 10 μm/min. DRIE exhibits excellent process controllability and creates the vias with a high aspect ratio up to 110:1. Depending on the etching process, a via-first or via-last process sequence is followed.

Via-first: From the front side of a full-thickness wafer, vias are etched during FEOL processing. Most of the logic suppliers have favored this approach. Using the via-first scheme, one can choose the smallest diameter and higher aspect ratio of 5 to 10 μm and 10:1, respectively.

Via-last: After completing the BEOL processing, vias are etched from the front side of a full-thickness wafer or backside of a thinned wafer. This approach has suitable applications for image sensors and stacked DRAM. The primary reason behind this approach is the minimum electrical resistance of vias. For instance, a CMOS image sensor exhibits via diameters exceeding 40 μm having aspect ratios of 2:1. In other devices, the via diameters are in the range of 10 to 25 μm with an aspect ratio of 5:1.

8.1.6.2   Deposition of Oxide

After completing the etching process, a trench is found in silicon. By filling the material with metal, silicon is diffused and the signal will get lost. The problem of diffusion can be avoided using an insulator of growing oxide in this trench. The major challenge behind this approach is the high aspect ratio of silicon. Therefore, the same thickness is required to impose at the top and bottom of the trench, as shown in Figure 8.1.

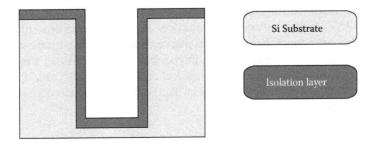

FIGURE 8.1   Deposition of isolation layer on silicon substrate.

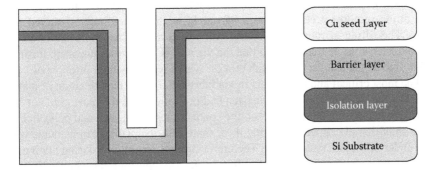

FIGURE 8.2   Grown process of Cu seed layer.

Oxide or nitride can be deposited at high temperatures like 900 or 1100°C. At a lower temperatures around 200 and 100°C, plasma-enhanced chemical vapor deposition (PECVD) and atomic layer deposition (ALD) are used to deposit the oxide or nitride layers, respectively [10]. These approaches provide good adhesion to silicon. While ALD oxide has low breakdown voltage and throughput, high-quality PECVD is desirable for TSV isolation.

For dielectric applications in TSV-based 3D interconnects, the chemical vapor deposition (CVD) is an ideal process. This process exhibits an inherently conformal deposition that is a critical coverage for subsequent titanium metal barrier and copper seed steps. To achieve a high aspect ratio for via-first and via-last TSV applications, a highly conformal deposition process is required. The deposited film exhibits good breakdown voltage and leakage current properties with excellent adhesion to the industry standard barrier metals (Ta, Ti, and TaN/Ta).

8.1.6.3   Barrier Layer or Seed Layer

A metal barrier layer or seed layer is placed to prevent the diffusion of metal into silicon or oxide after and before filling of oxide, as depicted in Figure 8.2. During the application of barrier layer, it is required to deposit the oxide to obtain similar thicknesses on both the top and the sides. The barrier metal should have good diffusion properties. In the process flow of TSV fabrication, the most challenging and expensive task is the deposition of barrier layer and subsequent filling of via.

The most favorable barrier materials are titanium and tantalum, which can be used for copper TSVs for their advanced logic devices. A physical vapor deposition (PVD) process is used to deposit these materials that can deliver highly uniform step coverage and sheet resistance. It results in a thinner barrier for depositing a metal layer with void-free, lower stress and good electrical conductivity [10]. Other approaches like ALD, CVD, and electroless plating can also be applied to produce a high-quality barrier/seed layer.

8.1.6.4   Via Filling/Plating

Tungsten is generally used to fill the via, which is a well-known process in the wafer fabrication community. In the 1980s, tungsten was first introduced to wafer fab processing to fill 1 μm diameter contacts and vias up to 2 μm deep [10]. The filling requires a contact and a barrier layer that is normally sputtered titanium. The filling of tungsten is conformal, which means that more than half the via diameter must be deposited so that the via will be filled with tungsten. It can be referred to as the standard wafer fab process with a relatively high throughput.

The most critical and costly part in via fabrication is the metallization or plating step. For the via-first approach, tungsten and polysilicon are the most used conductive fill materials for TSVs. However, the conductivity of tungsten and polysilicon is lower than that of copper. But, using these materials, the plating can be made void-free under minimum stress effect. Therefore, stress is minimized during plating.

The most used process for plating is electrodeposition of copper [11–14]. Electrodeposition is a well-known semiconductor process that is widely used to form a conducting path. This deposition process exhibits good processability and availability at room temperature. However, it suffers from several complexities, such as reliability, throughput, and process controllability [15]. Especially, high-aspect-ratio TSVs with void-free conductive TSV metal cores are difficult to implement. Therefore, an investigation is required to find some alternative approaches for plating processes. Some of these probable approaches are the use of solder balls [16], filling with conducive metal pastes [17], wire-bonded gold cores, etc.

8.1.6.5   Chemical Mechanical Polishing

After metallization, chemical mechanical polishing (CMP) is done to remove the undesirable oxide or metal. This method requires a two-phase polishing approach. The first phase uses a higher removal rate of oxide with good polarization and low nonuniformity. This phase does not concern the low dishing. The second phase follows the process with a lower removal rate and acceptable dishing that is usually selective to the barrier [18]. The different steps of CMP for a Cu-based TSV are presented in Figure 8.3. This polishing process suffers from several challenges that include rapid removal of thick materials without compromising wafer topography.

8.1.6.6   Wafer Thinning

The thinning process of wafers generally follows two steps: (1) deposition of metal films on the back to promote backside contact to the device and (2) reducing the TSV depth. Silicon substrates can be thinned after being bonded together to

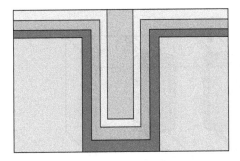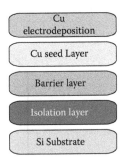

FIGURE 8.3   Cu-based TSV structure after CMP process.

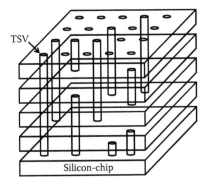

FIGURE 8.4   Through-silicon via chip stacking.

allow the interconnect to be formed on the back of the bonded wafer or to reveal an existing interconnect for bonding to another substrate.

Backgrinding is a generic process for thinning the wafer. In this process, a "backgrind tape" of approximate thickness of 100–300 μm is applied to the front to protect the front side of the device wafer [18]. A grinding wheel/disk with embedded diamonds is used to perform the grinding. The grinding is done in the following two steps: (1) a coarse grind to remove the bulk of the silicon and (2) a fine grind for thinning the wafer. For instance, a grit size of 50 μm is thinned up to 10–20 μm by using coarse grind, while it is reduced to 2 μm for fine grind [18]. TSVs are used to connect multiple 3D dies, as shown in Figure 8.4.

8.2   MODELING OF CU-BASED TSV

A Cu TSV can be modeled as a cylindrical conductor using copper metal and silicon substrate. An insulating layer is used to fill the gap between the Cu metal and the silicon substrate. Cu is preferred due to its economical feasibility and technical superiority. Furthermore, electrochemical deposition (ECD) technology can make Cu a suitable via material for higher electrical conductivity and thermal characteristics.

The physical configuration of the 3D model of a Cu-based TSV is shown in Figure 8.5(a). Depending on the configuration, an electrical equivalent circuit model

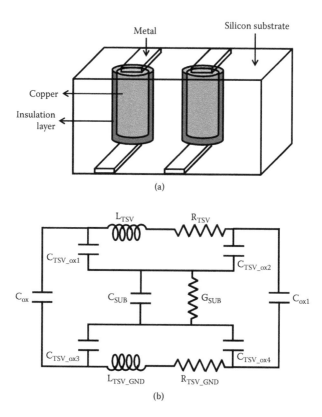

(a)

(b)

FIGURE 8.5   (a) Physical configuration and (b) equivalent electrical model of Cu-based TSV.

is presented in Figure 8.5(b). Since the model is derived from its realistic structure, it can physically represent all the parasitic components. The equivalent model of signal TSV primarily accounts for an inductor L_{TSV} and a resistor R_{TSV} that primarily depends on the TSV height (H_{TSV}), TSV diameter (D_{TSV}), pitch diameter (d_{pitch}), and resistivity (ρ_{TSV}) and can be expressed as [19]

$$R_{TSV} = \sqrt{(R_{DC})^2 + (R_{AC})^2}$$ (8.1)

where

$$R_{DC} = \rho_{TSV} \frac{H_{TSV}}{\pi(D_{TSV}/2)^2} \text{ and } R_{AC} = \rho_{TSV} \frac{H_{TSV}}{\pi D_{TSV} \delta_{skin}}$$ (8.2)

$$\delta_{skin} = \sqrt{\frac{1}{\sigma_{TSV} \pi \mu_0 \mu_r f}} \text{ and } L_{TSV} = \frac{\mu_0 \mu_r H_{TSV}}{2\pi} \ln\left(\frac{d_{pitch}}{D_{TSV}/2}\right)$$ (8.3)

In the equivalent model of Figure 8.5(b), C_{TSV_OX} represents the oxide capacitance between the via and the silicon, whereas C_{OX} represents the oxide capacitance and fringing capacitance between the vias. The C_{TSV_OX} primarily depends on the thickness of the SiO_2 layer (t_{ox}) and can be expressed as

$$C_{TSV\_OX} = 2\pi\varepsilon_0\varepsilon_{SiO_2} \frac{H_{TSV}}{\ln\left(\dfrac{D_{TSV}/2 + t_{ox}}{D_{TSV}/2}\right)} \tag{8.4}$$

Lossy characteristics of the silicon between the vias are represented by G_{SUB}, and the capacitance between the vias is characterized by C_{SUB}. They can be expressed as [19]

$$C_{SUB} = \pi\varepsilon_0\varepsilon_{Si} \frac{H_{TSV}}{\ln\left\{\dfrac{d_{pitch}}{D_{TSV}}\left[1 + \sqrt{1 - \left(\dfrac{D_{TSV}}{d_{pitch}}\right)^2}\right]\right\}} \tag{8.5}$$

and

$$G_{SUB} = \frac{C_{SUB}\sigma_{Si}}{\varepsilon_{Si}} \tag{8.6}$$

where σ_{Si} and ε_{Si} are the conductivity and permittivity of the silicon substrate, respectively.

8.2.1 EQUIVALENT ELECTRICAL MODEL FOR HIGH-FREQUENCY APPLICATIONS

TSV primarily provides the integration of analog, digital, radio frequency (RF), microelectromechanical systems (MEMS), etc., in a single package that reduces the space and results in high packaging density [20]. Therefore, Kannan et al. [21] developed an equivalent electrical model (Figure 8.6) that operates at 60 GHz frequencies. At higher frequencies, a skin effect emerges in Cu interconnects that results in higher current density on the surface than through the core. In order to develop the model, the frequency-dependent parasitics are incorporated in addition to

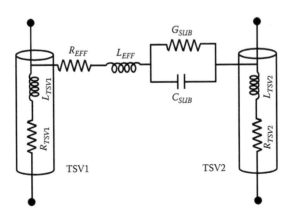

FIGURE 8.6   Equivalent electrical circuit model of Figure 8.5(a) for high-frequency applications.

the substrate parasitic. The modeling considers the substrate as a monolithic capacitor and skin effect of Cu at high frequencies. Therefore, the equivalent electrical circuit can model the substrate as a large parallel plate capacitor, whereas the associated parasitics can be modeled as R_{EFF} (electrostatic series resistance), L_{EFF} (electrostatic series inductance), capacitance, and conductance of the silicon substrate [22].

8.2.1.1 Effective Series Resistance (R_{EFF})

The resistance of TSVs and inner conductors can be modeled as effective series resistance (R_{EFF}). TSVs between two stacked dies are primarily considered transmission lines, and the R_{EFF} of the substrate is analyzed with the help of a resonant line technique. Thus, the series resistance of the substrate is given by [23, 24]

$$R = Zo\left[\frac{\pi}{4}\frac{f}{fo}Cosec^2\left(\frac{\pi}{4}\frac{f}{fo}\right) - \frac{1}{2}Cot\left(\frac{\pi}{4}\frac{f}{fo}\right)\right]\left(\frac{1}{Qs} - \frac{1}{Qo}\right) + \frac{\omega L}{Qs} \qquad (8.7)$$

A high Q-factor can be obtained using the resonant line that is modeled with the classical transmission line theory and can be expressed as

$$Q = 8.39b\sqrt{f} \qquad (8.8)$$

where f and b represent the frequency (in GHz) and inner radius of the TSV, respectively, by neglecting the effect of dielectric loss. All losses related to the skin effect of TSV need to be considered when the Q-factor of the system (Q_s) is smaller than the Q-factor of TSV. The Q-factor at the natural frequency (Q'_o) is used to obtain the Q-factor at the test frequency (Q_o) and can be expressed as

$$Qo \approx \left(\frac{f}{fo}\right)Qo' \qquad (8.9)$$

The current can exhibit its maximum and minimum values at odd and even harmonics, respectively. It may result in a resistive loss in the substrate that depends on the frequency component. Therefore, the resistive loss can be expressed as

$$R_0 = 0.004\left(\frac{f}{fo}\right)^{0.84} \qquad (8.10)$$

Combining Equations (8.7), (8.9), and (8.10), the R_{EFF} of the substrate can be obtained as

$$R_{EFF} = Zo\left[\frac{\pi}{4}\frac{f}{f_0}Cosec^2\left(\frac{\pi}{2}\frac{f}{f_0}\right) - \frac{1}{2}Cot\left(\frac{\pi}{2}\frac{f}{f_0}\right)\right]\left(\frac{1}{Qs} - \frac{1}{Qo\sqrt{\frac{f}{f_0}}}\right) + \frac{\omega L}{Qs} - R_0 \qquad (8.11)$$

For a test frequency and a natural frequency of 1.5 GHz and 1 GHz, respectively, the Q_s and Q'_o can be calculated as (8.4) and (8.9), respectively. Using the above-mentioned values along with $R_o = 7.2$ mΩ, the R_{EFF} is estimated as 5.69 mΩ/μm.

8.2.1.2 Effective Series Inductance (L_{EFF})

The effective series inductance (L_{EFF}) can be referred to as a function of TSV pitch and primarily depends on the self-inductance of the TSV. Thus, the L_{EFF} can be expressed as [25, 26]

$$L_{EFF} = L_{TSV}.K_g \tag{8.12}$$

where L_{TSV} and K_g are the self-inductance and the correction factor of TSV pitch, respectively. The correction factor primarily depends on the diameter of pitch and TSV and can be expressed as

$$K_g = K_{g\_a} - K_{g\_b} \cdot \ln\left(\frac{r_{via}}{t_{silicon} + t_{insulator}}\right) \tag{8.13}$$

where $r_{via} = D_{TSV}/2$, and $t_{silicon}$ and $t_{insulator}$ represent the thickness of the Si substrate and the insulating layer (SiO$_2$), respectively. The coefficients K_{g_a} and K_{g_b} can be determined using circuit optimization during the model extraction process for the silicon substrate and the insulating layer, respectively. For the coefficients K_{g_a} and K_{g_b} of 0.65 and 0.28, respectively, one can easily obtain the correction factor from (8.13). Equation (8.13) is further used to calculate the L_{EFF}.

8.2.1.3 Equivalent Parasitic Calculation

The parasitics associated with electrical equivalent models of Figure 8.5(b) and 8.6 primarily account for the dimensions of TSV, such as H_{TSV} and D_{TSV}, respectively. To avoid the electromigration effect at high frequency, a polymer can be used as an insulating layer between the metal TSV, whereas the silicon substrate is for the high-frequency applications. The thickness of the insulating layer and the diameter of TSV pitch are represented as $t_{insulator}$ and d_{pitch}, respectively. A TiN (titanium nitride) diffusion barrier with a thickness of $t_{diffusion}$ can be deposited to achieve the TSV metallization. The formulation of different parasitics associated with Cu-based TSV is discussed in this section.

8.2.1.3.1 Resistance

The self-resistance of Cu-based TSV can be expressed as [27]

$$R_{TSV} = R_{TiN} + \frac{H_{TSV}}{\sigma_{Cu} S_{skin}} \tag{8.14}$$

where σ_{Cu} and S_{skin} are the conductivity and the effective conductance of the Cu surface. Apart from this, R_{TiN} refers to the resistance of the diffusion barrier and can be expressed as

$$R_{TiN} = \frac{t_{diffusion}}{\sigma_{TiN} 4r_{via}^2} \tag{8.15}$$

where σ_{TiN} represents the conductivity of the diffusion barrier.

S_{skin} in Cu TSV is referred to as the tendency of the current distribution within a conductor such that the current density is larger near the surface than at the edge of the conductor. Thus, the S_{skin} can be expressed as

$$S_{skin} = 4(2r_{via}\,\delta - \delta^2) \tag{8.16}$$

where δ is the skin depth and can be expressed as

$$\delta = \sqrt{\frac{2\rho}{\omega\mu}} \tag{8.17}$$

ρ, ω, and μ are the resistivity, operating frequency, and relative permeability of Cu, respectively. The computation of total resistance in Cu-based TSV follows the expressions from (8.14) to (8.17).

8.2.1.3.2   Inductance

Self-inductance is defined as the property by which a change in current induces a voltage in the conductor. The self-inductance of TSV primarily depends on the pitch diameter and can be expressed as [28]

$$L_{via} = \frac{\mu}{\pi} \ln\left(\frac{d_{pitch}}{2r_{via}} + \sqrt{\left(\frac{d_{pitch}}{2r_{via}}\right)^2 - 1}\right) H_{TSV} \tag{8.18}$$

Generally, the self-inductance of the TSV is a function of frequency and can be expressed as

$$L_{TSV} = \frac{L_{via}}{1 + \log\left(\dfrac{f}{10^8}\right)^{0.26}} \tag{8.19}$$

8.2.1.3.3   Capacitance

The capacitance primarily defines the difference in potential between two charged bodies. Thus, the capacitance between the TSV and the silicon substrate can be expressed as

$$C_{SUB} = \frac{\varepsilon_0 \varepsilon_r A}{d_{pitch}} \tag{8.20}$$

where $A = \pi r_{via} H_{TSV}$ is the effective area considered for the capacitance.

8.2.1.3.4   Conductance

Electrical conductance is referred to as the inverse quantity of electrical resistance that can be defined as the ease with which the electric current passes through any material. Therefore, the electrical conductance of a silicon substrate can be expressed as [22]

$$G_{SUB} = \frac{\pi\sigma}{\ln\left(\frac{d_{pitch}}{2r_{via}} + \sqrt{\left(\frac{d_{pitch}}{2r_{via}}\right)^2 - 1}\right)} \quad (8.21)$$

where σ is the conductivity of the silicon substrate.

Using the expressions from (8.14) to (8.21), different parasitics such as effective resistance, inductance, and conductance can be calculated. Kannan et al. [21, 29] demonstrated the quantitative values of these parasitics for different via heights and diameters. It is explained that quantitative values of the parasitics are significantly influenced by the higher via dimensions.

8.2.2 LIMITATIONS OF CU-BASED TSVS

Although Cu-based TSV is an obvious choice, it still faces certain challenges that arise due to fabrication limitations in achieving proper physical vapor deposition (PVD), seed layer deposition for ECD, and performance limitations such as electromigration and higher resistivity. The resistivity can be attributed to a combined effect of scattering and the presence of a highly diffusive barrier layer that increases the difficulty in obtaining high-aspect-ratio vias [30]. Therefore, researchers are forced to find an alternative solution to replace the Cu-based TSVs. Carbon nanotubes (CNTs) are emerging as an interesting choice for filler material. Several modeling and simulations results [31] have been demonstrated using CNTs at relatively lower frequencies (~10 GHz). The observation reveals that CNTs have extraordinary potential to overcome the fundamental bottlenecks associated with present-day Cu TSVs.

Depending on the integrated model developed for low frequency, it can be observed that a single-walled CNT (SWCNT) bundle offers unique current-carrying capability ($>10^{10}$A/cm^2, nearly three orders of magnitude higher than that of Cu). The good thermal stability and negligible electromigration effects are the added advantages of CNT-based TSVs. Thus, compared to Cu TSVs, an impressive result can be obtained for CNT-based TSVs at higher frequencies.

8.3 MODELING OF CNT-BASED TSV

During the recent past, carbon nanotubes (CNTs) have emerged as a promising interconnect material for future miniaturized electronics [32]. Particularly, CNTs exhibit lower thermal expansion [33], joule heating [34], and electromigration [35], and higher mechanical stability, thermal stability, thermal conductivity, and current-carrying capability. Conventionally, these unique properties make CNTs an attractive choice as a reliable material for future very-large-scale integrated (VLSI) interconnects as well. Both the on-chip and off-chip CNT interconnects outperform the conventional Cu ones [36, 37]. The higher contact resistance of SWCNT is the primary challenge. Due to the reduced effect of metal-nanotube imperfect contact resistance, SWCNT bundles are often preferred over a single SWCNT as via material. Thus, the current research is targeted toward the development of CNT bundle-based vias and interconnects.

8.3.1 Theory and Modeling

The modeling of SWCNT bundle-based TSVs on a Si substrate primarily uses a bundled SWCNT as filler material. Silicon dioxide (SiO_2) is the most commonly used insulating material for TSVs. However, for high-frequency applications, SiO_2 cannot be used as an insulator due to its large fringing capacitance [38]. Hence, this insulating layer needs to be replaced with a suitable polymer layer. The capacitance of TSVs with polymer liners is reduced due to its lower dielectric constants and larger thickness than SiO_2 liners. It results in an improved electrical performance in terms of interconnect speed, power dissipation, and cross talk [39].

This section presents three different equivalent electrical models of SWCNT bundle-based TSVs. The equivalent models primarily represent the schematic of a pair of SWCNT bundle TSVs, as shown in Figure 8.7(a).

8.3.1.1 Model I [40]

Depending on the chirality, a SWCNT can exhibit either metallic or semiconducting properties. Performance of the pair of TSVs primarily depends on the number of metallic SWCNTs in the bundle (N_{CNT}), as shown in Figure 8.7(b). Thus, N_{CNT} can be expressed as [40]

$$N_{CNT} = \frac{2\pi r_{via\_CNT}^2}{\sqrt{3}\left(2r_{CNT} + d_{CNT}\right)^2} \tag{8.22}$$

where r_{CNT} and r_{via_CNT} are the radii of single and bundled SWCNTs, respectively. The modeling assumes that all the SWCNTs are closely packed. Depending on the Van der Waal's force between neighboring carbon atoms, each SWCNT in the

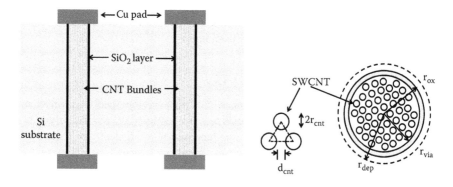

FIGURE 8.7 (a) Schematic of a pair of SWCNT bundle-based TSVs and (b) cross section of a SWCNT bundle.

bundle has a distance $(d_{CNT}) \approx 0.34$ nm. In order to derive the circuit model between a pair of SWCNT bundle-based TSVs, it is imperative to consider a complex effective conductivity that can be expressed as [40]

$$\sigma_{eff} = \frac{H_{TSV} N_{CNT}}{Z_{CNT} \pi r_{via\_CNT}^2} \quad (8.23)$$

where Z_{CNT} is the intrinsic self-impedance of a single SWCNT and can be determined as [40]

$$Z_{CNT} = \frac{R_{mc} + R_q + R_S + j\omega L_k}{N_{channel}}$$

$$= \frac{h}{2e^2 N_{channel}} \left(1 + \frac{H_{TSV}}{\lambda} + \frac{j\omega H_{TSV}}{2v_F} \right) \quad (8.24)$$

where R_q, R_s, and L_k are the quantum contact resistance, the scattering resistance, and the kinetic inductance of each conducting channel, respectively, and $v_F \approx 8 \times 10^5$ m/s is the Fermi velocity of CNT. λ and $N_{Channel}$ represent the mean free path of electrons and the total number of conducting channels, respectively. The imperfect contact resistance, R_{mc}, can be neglected, as it highly depends on the fabrication process. Due to the dominating value of kinetic inductance (L_k), this model neglects the effect of magnetic inductance.

An equivalent lumped model of a pair of SWCNT bundle TSVs is shown in Figure 8.8. Although the equivalent model considers the quantum effects, it neglects the coupling effects from metal contacts, the redistribution layer (RDL), or back-end-of-line (BEOL) to the TSVs. The above circuit model of Figure 8.8 is further simplified as the transmission line model, as presented in Figure 8.9.

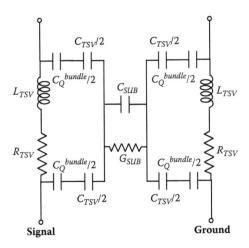

FIGURE 8.8   Equivalent electrical model of a pair of SWCNT bundle-based TSVs.

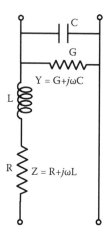

FIGURE 8.9   Simplified transmission line model of Figure 8.8.

The effective impedance of the pair of CNT-based TSVs can be determined as [40]

$$Z_{eff} = ZH_{TSV} = (R + j\omega L)H_{TSV} \tag{8.25}$$

The Z_{eff} depends on both the skin depths of TSV and the silicon substrate, δ_{tsv} and δ_{si}, respectively, and can be expressed as [40]

$$\delta_{tsv} = \sqrt{2/\omega\mu\sigma_{eff}} \tag{8.26}$$

$$\delta_{si} = \sqrt{2/\omega\mu(\sigma_{Si} + j\omega\varepsilon_{Si})} \tag{8.27}$$

where σ_{Si}, ε_{Si}, and μ represent the conductivity, permittivity, and permeability of silicon substrate, respectively.

The elements G_{SUB} and C_{SUB} in Figure 8.8 represent the conductance and capacitance of the silicon substrate, respectively, and can be expressed as [40]

$$C_{SUB} = \frac{\pi\varepsilon_{Si}H_{TSV}}{\cosh^{-1}\left(d_{pitch}/\left(2r_{dep}\right)\right)} \tag{8.28}$$

$$G_{SUB} = \sigma_{Si}C_{SUB}/\varepsilon_{Si} \tag{8.29}$$

The effective admittance of the pair of CNT TSVs can be expressed as

$$Y_{eff} = YH_{TSV} = \left(G + j\omega C\right)H_{TSV}$$

$$= \left[2\left(j\omega C_Q^{bundle}\right)^{-1} + 2\left(j\omega C_{TSV}\right)^{-1} + \left(G_{Si} + j\omega C_{Si}\right)^{-1}\right]^{-1} \tag{8.30}$$

where C_Q^{bundle} represents the equivalent quantum capacitance of the SWCNT bundle. C_{TSV} is the MOS capacitance in the inversion region and can be expressed as

$$C_{TSV} = \left(\frac{1}{C_{ox}} + \frac{1}{C_{dep}} \right)^{-1}$$

$$= \frac{\ln(r_{ox}/r_{via})}{2\pi\varepsilon_0 H_{TSV}} + \frac{\ln(r_{dep}/r_{ox})}{2\pi\varepsilon_{Si} H_{TSV}}$$

(8.31)

where r_{dep} and r_{ox} are the depletion and oxide radius, respectively, as shown in Figure 8.7(b). Physically, the TSV pair with higher electrical conductivity has lesser conductive loss and better transmission characteristics. The electrical conductivity of the SWCNT bundle is dependent on the frequency and the geometrical parameters.

8.3.1.2 Model II [31]

Modeling of a SWCNT bundle-based TSV integrates the equivalent electrical models of both TSV and SWCNT bundles. This type of integration is required to obtain an equivalent electrical model of TSV with a SWCNT bundle as the via material. Depending on the physical configuration of Figure 8.7(a), the equivalent electrical model of a SWCNT bundle-based TSV is shown in Figure 8.10. The model is compatible for RF application at 60 GHz frequencies [31].

Before moving to the descriptive modeling, it is required to understand the parasitics that effectively characterize the SWCNT and the substrate behavior. R_{EFF} and L_{EFF} represent the effective series resistance and inductance, respectively, in the equivalent model of Figure 8.10. Both the effective resistance and inductance are the frequency-dependent components, as discussed in Sections 8.2.1.1 and 8.2.1.2, respectively. L_{EFF} primarily depends on the TSV pitch, as presented in Equation (8.12). C_{SUB} and G_{SUB} represent the effective capacitance and conductance of Si substrate and can be expressed as Equations (8.28) and (8.29), respectively. The equivalent model of Figure 8.10 also includes the parasitic resistance (R_{TSV}) and inductance

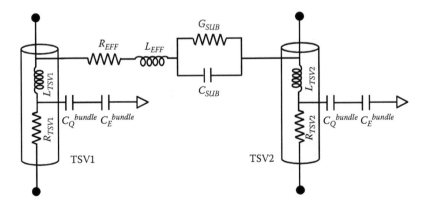

FIGURE 8.10 Equivalent electrical model of bundled SWCNT-based TSVs.

(L_{TSV}) of a SWCNT bundle and is discussed in Section 8.3.2. Via parasitics such as quantum capacitance (C_Q^{bundle}) and electrostatic capacitance (C_E^{bundle}) can also be taken into account that primarily depend on the number of SWCNTs in a bundle. Higher bundle density can be obtained with a higher number of SWCNTs in a bundle. For a higher number of SWCNTs, one can obtain more conducting channels in parallel, which effectively reduces the resistance to provide a better transmission and throughput.

8.3.1.3 Model III [29]

For RF applications (ranging from 2 to 20 GHz), Kannan et al. [29] presented an equivalent electrical model of SWCNT bundle TSV as shown in Figure 8.11. Between each via and the silicon substrate, SiO_2 is considered an insulating layer.

Each TSV in Figure 8.11 is electrically modeled as per unit length (*p.u.l.*) inductance (L'_{TSV}) and resistance (R'_{TSV}), and the parasitic capacitances between the vias.

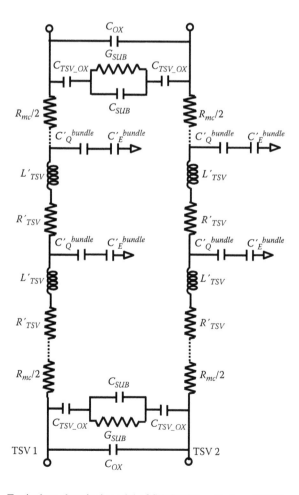

FIGURE 8.11 Equivalent electrical model of SWCNT bundle-based TSV.

Since SWCNT bundles are used as TSV filler material, R'_{TSV} and L'_{TSV} primarily represent the equivalent resistive and inductive parasitic of the SWCNT bundle, as discussed in Section 8.3.2. R_{mc} is the imperfect metal-nanotube contact resistance that primarily depends on the fabrication process. The equivalent capacitance of the bundle is categorized as (1) quantum capacitance (C'_{Q}^{bundle}) that primarily depends on the finite density of states of electrons and (2) electrostatic capacitance (C'_{E}^{bundle}) that occurs due to the potential difference between the SWCNT bundle and the ground plane [41]. C_{SUB} and G_{SUB} represent the p.u.l. capacitance and conductance of the Si substrate, respectively, and depend on the diameter of TSV pitch (d_{pitch}). Apart from this, C_{TSV_OX} and C_{OX} in Figure 8.11 primarily represent the capacitance of SiO_2 around a via and the fringing capacitance between two vias, respectively, and can be expressed as [42]

$$C_{TSV\_OX} = \frac{4\varepsilon_0\varepsilon_r H_{TSV}\left(D_{TSV}/2 - t_{ox}\right)}{t_{ox}} \tag{8.32}$$

$$C_{OX} = \left(\left(\frac{2}{C_{TSV\_OX}} + \left(\frac{\varepsilon_0\varepsilon_r A}{d}\right)^{-1}\right)^{-1}\right)^{-1} \tag{8.33}$$

8.3.2   PARASITIC FORMULATIONS FOR SINGLE AND BUNDLED SWCNTS

Modeling of CNT TSV primarily depends on the resistive, inductive, and capacitive parasitics of each SWCNT in the bundle. For a bundle, the total number of SWCNTs demonstrates the parasitic calculations.

8.3.2.1   Resistance Formulation

Nieuwoudt and Massoud [43] evaluated the resistance of individual SWCNTs in terms of the applied bias voltage. For nanotubes operating in a low-bias region ($V_b \leq 0.1$ V), the resistance is a summation of lumped quantum (R_q), contact (R_{mc}), and distributed per unit length scattering resistance (R_s).

$$R_{low} = R_q + R_{mc} if l_b \leq \lambda \tag{8.34}$$

$$R_{low} = R_s + R_{mc} if l_b \leq \lambda \tag{8.35}$$

where l_b is the length of nanotube and λ is the mean free path for acoustic phonon scattering [44]. For high-bias voltages ($V_b > 0.1$ V), the resistance of individual SWCNTs depends on the applied bias voltage:

$$R_{high} = R_{low} + \frac{V_b}{I_o} \tag{8.36}$$

where I_o is the maximum current flowing through an individual nanotube, approximately 20–25 μA [45].

8.3.2.1.1 Intrinsic and Contact Resistance

Burke [46] evaluated the conductance of a CNT using the two-terminal Landauer-Buttiker formulism as

$$G = \left(N_{channel} e^2 / h \right) \tag{8.37}$$

where $N_{channel}$ is the number of conducting channels in parallel, h is Planck's constant, and e is the charge of a single electron. From the band structure [47], it is observed that CNTs have two modes of propagation due to their twofold band degeneracy. Further, in each mode of propagation, the electrons can be either spin up or spin down. This results in a total of four conducting channels in SWCNT. Therefore, the quantum resistance (R_q) of a ballistic SWCNT can be expressed as

$$R_q = \frac{h}{4e^2} \approx 6.5 K\Omega \tag{8.38}$$

An individual SWCNT will have a minimum resistance R_q, regardless of its length. R_{mc} models the higher lumped resistance due to imperfect metal-nanotube contacts. With the improvement in fabrication and bonding techniques, the additional imperfect metal-nanotube contact resistance is significantly reduced. Liang et al. [19] experimentally showed that R_{mc} approaches zero [20], and therefore

$$R_q + R_{mc} = R_q \tag{8.39}$$

R_{mc} is also dependent on the diameter of the nanotube. The R_{mc} of a SWCNT predominantly increases when the diameter of nanotube (d_t) is less than 1.0 nm [48]. Based on the experimental results obtained by Kim et al. [48], one empirical parameter, D_{rc}, is defined that can be modified on the basis of different fabrication techniques, contact materials, and contact bonding configurations. Based on the experimental data, R_{mc} remains relatively constant near the nominal value R_{cnom} for nanotubes with diameters greater than 2.0 nm. Therefore, the overall contact resistance can be modeled as [43, 49]

$$R_{mc} = D_{rc} R_{cnom} \ if \ 1.0 \le d_t \le 2.0 \ nm \tag{8.40}$$

$$R_q = R_{cnom} \ if \ d_q > 2.0 \ nm \tag{8.41}$$

Using Equations (8.40) and (8.41), it can be shown that nanotubes with smaller diameters can significantly increase the contact resistance.

8.3.2.1.2 Scattering Resistance

At low-bias voltage, the scattering resistance (R_s) of SWCNT can be expressed as

$$R_s = \frac{h}{4e^2} \frac{l_b}{\lambda} \tag{8.42}$$

The experimental results in [50, 51] demonstrated that λ primarily depends on the diameter of SWCNT. The resistance of an individual SWCNT versus diameter is governed by the following relation:

$$R_s = \frac{h\alpha l_b T}{4e^2 v_F d_t}$$

(8.43)

where v_F is the Fermi velocity in graphene ($\approx 8 \times 10^5$ m/s), T is the temperature in Kelvin, and α is the total scattering rate of the SWCNT. From (8.43), it is clear that ohmic resistance is inversely proportional to the diameter of the nanotube. In the fabrication process, the diameter of SWCNTs can be controlled, which results in reduced ohmic resistance.

The maximum conductance associated with CNT, taking into account spin degeneracy and sublattice degeneracy of electrons in graphene, is given by $4e^2/h = 155$ μS (as in the Landauer-Buttiker formulism) [52]. When the length of CNT exceeds the typical mean free path, additional ohmic resistance due to scattering has to be taken into consideration and is given by $(h/4e^2)$ L/λ, where L is the length of CNT.

8.3.2.2 Inductance Formulation

Inductance is defined as the resistance to current change due to Faraday's law, and represents the energy stored in the magnetic field generated by current, $\frac{1}{2}L_M I^2$ [47]. Here, L_M is the self-inductance and I is the current flowing through the conductor. On the other hand, mass is the mechanical counterpart for inductance, as it opposes the velocity change and gives rise to kinetic energy, $\frac{1}{2}mv^2$. When this kinetic energy of electrons is comparable to or larger than the stored energy in a magnetic field, the kinetic inductance needs to represent the kinetic energy of electrons. From the classical model of the current equation [53], the kinetic inductance is inversely proportional to the cross-sectional area of a conductor. As a result, when the cross-sectional area decreases, fewer carriers will be available and electrons will have to move faster to maintain a constant current. On the other hand, the magnetic inductance of a conductor (or wire) depends mainly on its distance to a return path and weakly on the cross-sectional area. Thus, as the cross-sectional dimension of a conductor (case of CNT) becomes smaller, the ratio of kinetic inductance to magnetic inductance becomes larger.

8.3.2.2.1 Magnetic Inductance

The magnetic inductance per unit length in the presence of a ground plane can be expressed as [46]

$$L_M^{CNT} = \frac{\mu}{2\pi} \cosh^{-1}\left(\frac{2y}{d}\right) \approx \frac{\mu}{2\pi} \ln\left(\frac{y}{d}\right)$$

(8.44)

where d is the nanometer diameter and y is the distance to the ground plane. The magnetic inductance is calculated from a magnetic field of an isolated current-carrying wire with some distance y from the ground plane, as shown in Figure 8.12.

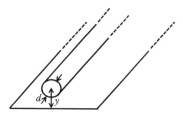

FIGURE 8.12   Geometry of nanotube in the presence of a ground plane.

In this case, for $d = 1$ nm and $y = 1$ μm, the calculated value of L_M^{CNT} (per unit length) is 1.4 pH/μm.

8.3.2.2.2 Kinetic Inductance

In order to calculate the kinetic inductance per unit length, Burke [46] estimated kinetic energy per unit length and equated that energy was $\frac{1}{2}LI^2$ of the kinetic inductance. The kinetic energy per unit length in a 1D wire (case of CNT) can be defined as the sum of the kinetic energies of left-movers and right-movers. For zero current, the numbers of electrons moving from left to right and right to left are the same, and as a result, they cancel the impact of one another, as shown in Figure 8.13(a). In order to generate a current from left to right, some of the left-movers have to convert to right-movers, as shown in Figure 8.13(b). By Pauli's exclusion principle, the converted electrons have to fill the higher energy levels. If the Fermi level of the left-movers is increased by $e\Delta\mu/2$, then the Fermi level of the right movers is reduced by the same amount, $e\Delta\mu/2$. Thus, the resulting current in 1D wire is

$$I = \frac{e^2}{h\Delta\mu} \tag{8.45}$$

The net increase in energy of the system is obtained by adding $e\Delta\mu/2$ energy per electron, where the excess number of electrons is given by $N = e\Delta\mu/2\Delta$. Here, Δ is the single particle energy level spacing and is related to the Fermi velocity v_F by

$$\Delta = \hbar v_F 2\pi/L \tag{8.46}$$

where $\hbar$ is known as Dirac's constant and is given by $h/2\pi$.

Therefore, the excess kinetic energy is given by $hI^2/4v_F e^2$. By equating this energy with $\frac{1}{2}LI^2$, the following expression can be obtained for kinetic energy per unit length for each conducting channel:

$$L_k = \frac{h}{2e^2 v_F} \tag{8.47}$$

The Fermi velocity of graphene and CNT, $v_F = 8 \times 10^5$ m/s, and therefore the kinetic inductance (per unit length) of each conducting channel, can be expressed as $L_k = 16nH/$μm. Individual SWCNTs have two propagating channels due to their band

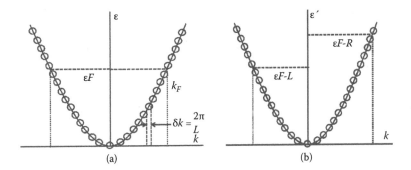

FIGURE 8.13 Electron energy versus wave-vector in a quantum wire. (a) For zero current, the numbers of right-mover electrons (positive k) and left-mover electrons (negative k) are equal. (b) Current flowing from left to right.

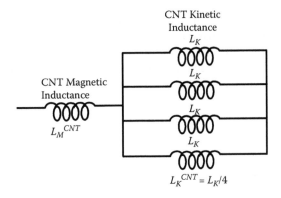

FIGURE 8.14 Equivalent inductance of SWCNT.

structure [46]. Each propagating channel is associated with spin up and spin down of electrons that result in the total number of four conducting channels per nanotube in parallel, as shown in Figure 8.14. Thus, the equivalent kinetic inductance for a SWCNT can be expressed as [49]

$$L_k^{CNT} = \frac{L_k}{4} \approx 4nH/\mu m \tag{8.48}$$

8.3.2.3 Capacitance Formulation

For a one-dimensional nanotube structure, the total capacitance depends on both the geometry and density of states [47]. As CNT is a one-dimensional wire, the total capacitance (C_{Total}) of a SWCNT is governed by electrostatic capacitance (C_E^{CNT}, geometric capacitance) and quantum capacitance (C_Q^{CNT}, density of states). Therefore, the total capacitance of CNT can be expressed as

$$\frac{1}{C_{Total}} = \frac{1}{C_E^{CNT}} + \frac{1}{C_Q^{CNT}} \tag{8.49}$$

8.3.2.3.1 Electrostatic Capacitance

Electrostatic capacitance can be modeled by considering SWCNT as a thin wire, with a diameter d placed at a distance y from the ground plane, as shown in Figure 8.12 [46, 54]. Thus, the electrostatic capacitance (C_E^{CNT}) between the SWCNT wire and the ground plane can be expressed as

$$C_E^{CNT} = \frac{2\pi\varepsilon}{\cosh^{-1}\left(\frac{2y}{d}\right)} \approx \frac{2\pi\varepsilon}{\ln\left(\frac{y}{d}\right)} \tag{8.50}$$

where ε is the permittivity. For $d = 1$ nm and $y = 1$ μm, the C_E^{CNT} (per unit length) is estimated as 30 aF/μm.

8.3.2.3.2 Quantum Capacitance

Quantum mechanics [55] states that no energy is required to add an electron in a classical electron gas. This implies that it is possible to add an electron with any arbitrary energy to the system. In a quantum electron gas, it is impossible to add an electron with energy less than the Fermi energy, E_F, due to Pauli's exclusion principle. In this case, it is possible to add an electron at an available quantum state above E_F. In a 1D system (CNT of length L), the spacing between quantum states can be expressed as

$$\delta E = \frac{dE}{dk}\delta k = \hbar v_F \frac{2\pi}{L} \tag{8.51}$$

Thus, the quantum capacitance per unit length of each conducting channel can be expressed as [56]

$$C_Q = \frac{2e^2}{hv_F} \tag{8.52}$$

As SWCNT has four conducting channels in parallel, the effective quantum capacitance is given by $C_Q^{CNT} = 4C_Q$ [57], as shown in Figure 8.15.

8.3.3 Limitations of CNT-Based TSVs

Although SWCNT bundle-based TSV demonstrates improved electrical, mechanical, and thermal behavior over Cu-based TSV, it faces some challenges due to its high imperfect metal-nanotube contact resistance [58–60]. The contact resistance primarily depends on the diameter of CNTs during fabrication growth. It is not trivial to control the growth of SWCNT with smaller diameters that effectively increase the overall contact resistance.

Fabrication of a SWCNT bundle-based TSV is an extremely tedious process. Certain issues like chirality of SWCNT, orientation, and density of bundles are the major challenges for developing SWCNT bundles as filler material [31, 61]. Due to these challenges, the fabrication techniques are limited to CVD and laser beam manipulation techniques. While growing vertical SWCNTs by CVD, the impinging ions can have a major effect on the substrate that leads to serious difficulties.

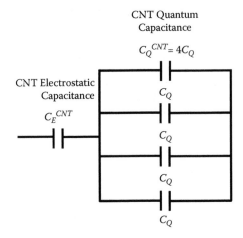

FIGURE 8.15   Equivalent capacitance of SWCNT.

The higher growth temperature required during fabrication is another major limitation to the application of CNT vias. Temperatures greater than 600°C are generally employed for carrying out CNT growth [58, 62], which is unfortunately incompatible with CMOS devices and many other temperature-sensitive materials.

8.4   SUMMARY AND CONCLUSION

This chapter presented the fabrication and modeling approach for Cu- and CNT-based through-silicon vias (TSVs). Initially, the fabrication and modeling of Cu-based TSVs along with their limitations were discussed. The main drawback of Cu-based TSV is the rapid increase in resistivity under the combined effects of enhanced grain boundary scattering, surface scattering, and the presence of a highly diffusive barrier layer. These limitations can be removed using CNTs that result in effective via designing. CNTs can exhibit higher mechanical and thermal stability, higher conductivity, and larger current-carrying capability. Moreover, a bundle of CNTs conducts current in a parallel fashion that effectively reduces the resistivity. Thus, bundled CNTs can be predicted as one of the potential candidates for future high-speed TSVs.

Based on the geometry and physical configurations, different electrical equivalent models for Cu- and SWCNT bundle-based TSVs are presented. The substrate and interconnect parasitics in the equivalent model are a function of frequency. For high-frequency applications, substrate-dependent conductance and capacitance have also been incorporated.

REFERENCES

1. Banerjee, K., Souri, S. J., Kapur, P., Saraswat, K. C. 3D ICs: a novel chip design for improving deep-submicrometer interconnect performance and systems-on-chip integration. *Proceedings of the IEEE*, 89(5): 602–633 (2001).

2. Davis, W. R., Wilson, J., Mick, S., Xu, J., Hao, H., Mineo, C., Sule, A. M., Steer, M., Franzon, P. D. Demystifying 3D ICs: the pros and cons of going vertical. *IEEE Design and Test of Computers*, 22(6): 498–510 (2005).

3. Souri, S. J., Banerjee, K., Mehrotra, A., Saraswat, K. C. Multiple Si layer ICs: motivation, performance analysis and design implications. In *Proceedings of IEEE 37th Conference on Design Automation (DAC '00)*, 213–220 (2000).

4. Reber, M., Tielert, R. Benefits of vertically stacked integrated circuits for sequential logic. In *Proceedings of the IEEE International Symposium on Circuits and Systems*, 4: 121–124 (1996).

5. Akasaka, Y. Three-dimensional IC trends. *Proceedings of IEEE*, 74: 1703–1714 (1986).

6. Morrow, P. R., Park, C.-M., Ramanathan, S., Kobrinsky, M. J., Harmes, M. Three-dimensional wafer stacking via Cu-Cu bonding integrated with 65-nm strained-Si/low-k CMOS technology. *IEEE Electron Device Letters*, 27(5): 335–337 (2006).

7. Jacob, P., Erdogan, O., Zia, A., Belemjian, P. M., Kraft, R. P., McDonald, J. F. Predicting the performance of a 3D processor-memory chip stack. *IEEE Design and Test of Computers*, 22: 540–547 (2005).

8. Alam, S. M., Troxel, D., Thompson, C. V. A comprehensive layout methodology and layout-specific circuit analyses for three-dimensional integrated circuits. In *Proceedings of IEEE International Symposium on Quality Electronic Design (ISQED 2002)*, 246–251 (2002).

9. Lu, K. H., Zhang, X., Ryu, S. K., Im, J., Huang, R., Ho, P. S. Thermo-mechanical reliability of 3-D ICs containing through silicon vias. In *Proceedings of IEEE Electronic Components and Technology Conference (ECTC)*, 630–634 (2009).

10. Xu, Z., Lu, J.-Q. Through-silicon-via fabrication technologies, passives extraction, and electrical modeling for 3-D integration/packaging. *IEEE Transactions on Semiconductor Manufacturing*, 26(1), 23–34 (2013).

11. Nilsson, P., Ljunggren, A., Thorslund, R., Hagstrom, M., Lindskog, V. Novel through-silicon via technique for 2d/3d SiP and interposer in low-resistance applications. In *Proceedings of IEEE 59th Electronic Components and Technology Components (ECTC '09)*, 1796–1801 (2009).

12. Wolf, M. J., Dretschkow, T., Wunderle, B., Jurgensen, N., Engelmann, G., Ehrmann, O., Uhlig, A., Michel, B., Reichl, H. High aspect ratio TSV copper filling with different seed layers. In *Proceedings of IEEE 58th Electronic Components and Technology Components (ECTC '08)*, 563–570 (2008).

13. Koyanagi, M., Fukushima, T., Tanaka, T. High-density through silicon vias for 3-D LSIs. In *Proceedings of the IEEE*, 97(1), 49–59 (2009).

14. Tezcan, D. S., Pham, N., Majeed, B., De Moor, P., Ruythooren, W., Baert, K. Sloped through wafer vias for 3D wafer level packaging. In *Proceedings of IEEE Electronic Components and Technology Conference (ECTC)*, 643–647 (2007).

15. Garrou, P., Bower, C., Ramm, P. *Handbook of 3D Integration Technology and Application of 3D Integration Circuits*. Wiley, 153 (2008).

16. Gu, J., Pike, W. T., Karl, W. J. A novel vertical solder pump structure for through-wafer interconnects. In *Proceedings of IEEE 23rd International Conference on Micro Electro Mechanical Systems (MEMS 2010)*, 500–503 (2010).

17. Motoyoshi, M. Through-silicon via (TSV). In *Proceedings of the IEEE*, 97(1), 43–48 (2009).

18. Hosali, S., Smith, G., Smith, L., Vitkavage, S., Arkalgud, S. Through-silicon via fabrication, backgrind, and handle wafer technologies. In C. S. Tan, R. S. Gutmann, and L. R. Reif (Eds.), *Wafer Level 3-D ICs Process Technology Integrated Circuits and Systems*. Berlin: Springer, 1–32 (2008).

19. Liang, L., Miao, M., Li, Z., Xu, S., Zhang, Y., Zhang, X. 3D modeling and electrical characteristics of through-silicon-via (TSV) in 3D integrated circuits. In *Proceedings of IEEE 12th International Conference on Electronic Packaging Technology and High Density Packaging (ICEPT-HDP)*, Shanghai, 1–5 (2011).

20. Liang, W., Bockrath, M., Bozovic, D., Hafner, J. H., Park, H. Fabry Perot interference in a nanotube electron waveguide. *Nature*, 411(6838): 665–668 (2001).
21. Kannan, S., Evana, S. S., Gupta, A., Kim, B., Li, L. 3-D copper-based TSV for 60-GHz applications. In *Proceedings of IEEE 61st Electronic Components and Technology Conference (ECTC)*, Lake Buena Vista, FL, 1168–1175 (2011).
22. Bermond, C., Cadix, L., Farcy, A., Lacrevaz, T., Leduc, P., Flechet, B., High frequency characterization and modeling of high density TSV in 3D integrated circuits. In *13th Workshop on Signal Propagation on Interconnects*, 1–4 (2009).
23. Lakshminarayanan, B., Gordon, H. C., Jr., Weller, T. M. A substrate-dependent CAD model for ceramic multilayer capacitors. *IEEE Transactions on Microwave Theory and Techniques*, 48(10): 1687–1693 (2000).
24. Maher, J. P., Jacobsen, R. T., Laferty, R. E. High-frequency measurements of Q-factors of ceramic chip capacitors. *IEEE Transactions on Components, Hybrids and Manufacturing Technology*, 1(3): 257–264 (1978).
25. Pucel, R. A. Design considerations for monolithic microwave circuits. *IEEE Transactions on Microwave Theory Technology*, 29(6): 513–534 (1981).
26. Benabe, E. Automated characterization of ceramic multilayer capacitors. In *Proceedings of IEEE 52nd Advanced RF Techniques Group Conference*, 88–94 (1998).
27. Ryu, C., Lee, J., Lee, H., Lee, K., Oh, T., Kim, J. High frequency electrical model of through wafer via for 3-D stacked chip packaging. In *Proceedings of IEEE Electronics System Integration Technology Conference*, 1, 215–220 (2006).
28. Kikuchi, H., Yamada, Y., Ali, A. M., Liang, J., Fukushima, T., Tanaka, T., Koyanagi, M. Tungsten through-silicon via technology for three dimensional LSIs. *Japanese Journal of Applied Physics*, 47, 2801–2806 (2008).
29. Kannan, S., Gupta, A., Kim, B. C., Mohammed, F., Ahn, B. Analysis of carbon naotube based through silicon vias. In *Proceedings of IEEE 60th Electronic Components and Technology Conference (ECTC 2010)*, 51–57 (2010).
30. Steinhogl, W., Schindler, G., Steinlesberger, G., Traving, M., Engelhardt, M. Comprehensive study of the resistivity of copper wires with lateral dimensions 100 nm and smaller. *Journal of Applied Physics*, 97(2): 023706-1–023706-7 (2005).
31. Gupta, A., Kim, B. C., Kannan, S., Evana, S. S., Li, L. Analysis of CNT based 3D TSV for emerging RF applications. *Proceedings of IEEE 61st Electronic Components and Technology Conference (ECTC)*, Lake Buena Vista, FL, 2056–2059 (2011).
32. Banerjee, K., Srivastava, N., Xu, C., Li, H. Carbon nanomaterials for next-generation interconnects and passives: Physics, status, and prospects. *IEEE Transactions Electron Devices*, 56(9), 1799–1821 (2009).
33. Jiang, H., Liu, B., Huang, Y., Hwang, K. C. Thermal expansion of single wall carbon nanotubes. *Journal of Engineering Materials and Technology*, 126(3): 265–270 (2004).
34. Ragab, T., Basaran, C. Joule heating in single-walled carbon nanotubes. *Journal of Applied Physics*, 106(6), 063705-1–063705-5 (2009).
35. Collins, P. G., Hersam, M., Arnold, M., Martel, R., Avouris, P. Current saturation and electrical breakdown in multiwalled carbon nanotubes. *Physical Review Letters*, 86(14): 3128–3131 (2001).
36. Naeemi, A., Meindl, J. D. Compact physical models for multiwall carbon-nanotube interconnects. *IEEE Electron Device Letters*, 27(5): 338–340 (2006).
37. Kim, B. C., Kannan, S., Gupta, A., Mohammed, F., Ahn, B. Development of carbon nanotube based through-silicon vias. *Journal of Nanotechnology in Engineering and Medicine*, 1(2): 021012-1–021012-8 (2010).
38. Huang, C., Chen, Q., Wu, D., Wang, Z. High aspect ratio and low capacitance through-silicon-vias (TSVs) with polymer insulation layers. *Microelectronics Engineering*, 104, 12–17 (2013).

39. Tezcan, D. S., Duval, F., Philipsen, H., Luhn, O., Soussan, P., Swinnen, B. Scalable through silicon via with polymer deep trench isolation for 3D wafer level packaging. In *Proceedings of IEEE Electronic Components and Technology Conference (ECTC)*, 1159–1164 (2009).

40. Zhao, W.-S., Yin, W.-Y., Guo, Y.-X. Electromagnetic compatibility-oriented study on through silicon single-walled carbon nanotube bundle via (TS-SWCNTBV) arrays. *IEEE Transactions on Electromagnetic Compatibility*, 54(1): 149–157 (2012).

41. Gupta, A., Kannan, S., Kim, B. C., Mohammed, F., Ahn, B. Development of novel carbon nanotube TSV technology. In *Proceedings of IEEE 60th Electronic Components and Technology Conference (ECTC 2010)*, 1699–1702 (2010).

42. Kim, B., Gupta, A., Kannan, S., Md, F., Ahn, B. Method and model of carbon nanotube based through silicon vias (TSV) for RF applications. U.S. Patent 2012/0306096 A1 (2012).

43. Nieuwoudt, A., Massoud, Y. Evaluating the impact of resistance in carbon nanotube bundles for VLSI interconnect using diameter-dependent modeling techniques. *IEEE Transactions Electron Devices*, 53(10): 2460–2466 (2006).

44. Park, J. Y., Rosenblatt, S., Yaish, Y., Sazonova, V., Ustunel, H., Braig, S., Arias, T. A., Brouwer, P. W., McEuen, P. L. Electron–phonon scattering in metallic single-walled carbon nanotubes. *Nano Letters*, 4(3): 517–520 (2004).

45. Yao, Z., Kane, C. L., Dekker, C. High-field electrical transport in single-wall carbon nanotubes. *Physical Review Letters*, 84(13): 2941–2944 (2000).

46. Burke, P. J. Luttinger liquid theory as a model of the gigahertz electrical properties of carbon nanotube. *IEEE Transactions on Nanotechnology*, 1(3): 129–144 (2002).

47. Javey, A., Kong, J. *Carbon nanotube electronics*. Berlin: Springer (2009).

48. Kim, W., Javey, A., Tu, R., Cao, J., Wang, Q., Dai, H. Electrical contacts to carbon nanotubes down to 1 nm in diameter. *Applied Physics Letters*, 87(17): 173101-1–173101-3 (2005).

49. Nieuwoudt, A., Massoud, Y. Understanding the impact of inductance in carbon nanotube bundles for VLSI interconnect using scalable modeling techniques. *IEEE Transactions on Nanotechnology*, 5(6): 758–765 (2006).

50. Zhou, X., Park, J. Y., Huang, S., Liu, J., McEuen, P. L. Band structure, phonon scattering, and the performance limit of single-walled carbon nanotube transistors. *Physical Review Letters*, 95(14): 146805-1–146805-3 (2005).

51. Pennington, G., Goldsman, N. Low-field semi-classical carrier transport in semiconducting carbon nanotubes. *Physical Review B, Condensed Matter*, 71(20): 205318-1–205318-12 (2005).

52. Datta, S. *Electronic transport in mesoscopic systems*. Cambridge: Cambridge University Press (1995).

53. Pond, J. M., Claassen J. H., Carter, W. L. Measurement and modeling of kinetic inductance microstrip delay lines. *IEEE Transactions on Microwave Theory and Techniques*, MTT-35, 1256–1262 (1987).

54. Li, H., Xu, C., Srivastava, N., Banerjee, K. Carbon nanomaterials for next-generation interconnects and passives: physics, status and prospects. *IEEE Transactions Electron Devices*, 56(9): 1799–1821 (2009).

55. Griffiths, D. J. *Introduction to quantum mechanics*, 2nd ed. Upper Saddle River, NJ: Pearson Prentice Hall (2004).

56. Bockrath, M. W. Carbon nanotubes: electrons in one dimension. PhD dissertation, University of California, Berkeley, CA (1999).

57. Banerjee, K., Im, S., Srivastava, N. Can carbon nanotubes extend the lifetime of on-chip electrical interconnections? In *Proceedings of IEEE 1st International Conference on Nano-Networks and Workshops*, 1–9 (2006).

58. Hwang, S. H., Kim, B.-J., Lee, H.-Y., Joo, Y.-C. Electrical and mechanical properties of through-silicon-vias and bonding layers in stacked wafers for 3D integrated circuits. *Journal of Electronic Materials*, 41(2): 232–240 (2012).

59. Fischer, A. C., Roxhed, N., Stemme, G., Niklaus, F. Low-cost through silicon vias (TSVs) with wire-bonded metal cores and low capacitive substrate-coupling. In *Proceedings of IEEE 23rd International Conference on Micro Electro Mechanical Systems (MEMS 2010)*, 480–483 (2010).

60. Puretzky, A. A., Geohegan, D. B., Jesse, S., Ivanov, I. N., Eres, G. In situ measurements and modeling of carbon nanotube array growth kinetics during chemical vapor deposition. *Applied Physics A, Materials Science and Processing*, 81(2): 223–240 (2005).

61. Rimskog, M. Through wafer via technology for MEMS and 3D integration. In *Proceedings 32nd IEEE/CPMT International Electronic Manufacturing Technology Symposium 2007 (IEMT '07)*, 286–289 (2007).

62. Yu, Z., Burke, P. J. Microwave transport in metallic single-walled carbon nanotubes. *Nano Letters*, 5(7): 1403–1406 (2005).

9 Low-Power Testing for 2D/3D Devices and Systems

Xijiang Lin, Xiaoqing Wen, and Dong Xiang

CONTENTS

9.1   INTRODUCTION

Power dissipation has become a primary concern in large-scale integration (LSI) design [1]. Over the years, LSI designers have been tirelessly developing hardware and software techniques for effectively reducing *functional power* [2]. Functional constraints are often exploited at various levels to remove wasteful operations so that functional power is reduced. For example, a circuit designed for cell phone applications contains many functional blocks that do not need to be activated simultaneously, e.g., a block for audio and a block for phone calls. Dynamically disabling unnecessary circuit blocks significantly reduces functional power.

However, power concerns do not end with the tape-out of a low-power design. In reality, each fabricated low-power circuit needs to be tested for manufacturing defects by applying test stimuli to the circuit, obtaining its responses, and comparing them with expected responses to determine whether the circuit is defective. Obviously, power dissipation also occurs during testing, and *test power* has been known to be much higher than functional power [3, 4]. One major reason is that multiple circuit blocks are often activated simultaneously for maximizing test efficiency in order to meet strict test cost requirements. Excessive test power often causes heat damage or circuit malfunction, leading to undue test-induced yield loss and reliability degradation [5, 6].

To tackle this serious problem of excessive test power, effective and efficient low-power testing must be conducted through sophisticated test power analysis and an optimal combination of test power reduction techniques. The objective of this chapter is to provide readers with basic information about LSI testing and to familiarize them with state-of-the-art low-power testing solutions.

9.2   BASICS OF LSI TESTING

The purpose of *LSI testing* is to determine whether an LSI circuit is defect-free [7–9]. It is conducted by (1) applying test stimuli to the *circuit-under-test* (CUT), (2) measuring its actual responses, and (3) comparing them with its expected responses to make GO/NG calls.

LSI testing has two primary goals: *high test quality* and *low test costs*. Test quality is mainly determined by the ratio of *undertest* (i.e., defective circuits passing a test) and the ratio of *overtest* (i.e., defect-free circuits failing a test). Undertest is mostly caused by inadequate test stimuli or *test patterns*. Overtest is conventionally blamed on the fact that some functionally benign defects may fail a circuit during testing. On the other hand, test costs are mainly determined by test data volume and test time. Different from design and manufacturing, where reuse and parallelism significantly reduce costs due to well-scaled efficiency, testing basically needs to deal with all individual components (transistors, interconnects, etc.) in the CUT, and usually does not benefit from improved efficiency in design and manufacturing. As a result, test costs have kept increasing over the years, and may eventually become higher than manufacturing costs.

LSI testing technologies have been evolving around these two goals, especially in the fields of *test generation* and *design for testability* (DFT). The former is about how to determine the contents of test patterns so as to achieve higher test quality with fewer test patterns, while the latter is about how to modify a circuit design so as to make its testing easier (e.g., making it possible to conduct test generation and test application in practical time) or more efficient (e.g., making it possible to reduce test data or test time).

Recently, power dissipation during testing, i.e., *test power*, has emerged as a new threat to the quality and costs of LSI testing, especially for low-power circuits, since test power can be much higher than functional power [10–12]. Excessive test power may cause heat damage and circuit malfunction. If left uncontained, test power may compromise test quality due to overtest; if inefficiently managed, test costs may get out of control. This means that *low-power testing* is indispensable to modern low-power LSI circuits.

9.2.1 Basics of Test Generation

Test generation is the process of creating test patterns for a CUT [7–9]. Its goal is to determine whether the CUT is free of any manufacturing defects by applying test patterns and comparing measured test responses against expected test responses. In the following, three basic concepts in test generation are briefly described.

9.2.1.1 Fault Model

Although test patterns are intended for determining whether a CUT is defect-free, it is impossible to enumerate all physical defects as direct targets in test generation. In practice, a *fault model* assumed in a certain circuit model is used to represent the logic behaviors of physical defects. Some typical fault models are as follows:

Stuck-at fault model: A stuck-at fault assumes that the logic value of a signal line in a circuit is fixed at either 0 (*stuck-at-0*) or 1 (*stuck-at-1*). Although intuitively a short defect to the ground (power supply) corresponds to a stuck-at-0 (stuck-at-1) fault, many other complex structural defects are also be represented by stuck-at faults.

Bridging fault model: A bridging fault assumes that two or more signal lines are shorted together and produce a final logic value according to a certain rule. It provides a potentially better representation for short-type structural defects, although the final logic value is often hard to determine.

Transition delay fault model: A transition delay fault assumes a transition ($0 \rightarrow 1$ or $1 \rightarrow 0$) at a signal line cannot propagate to circuit outputs fast enough with respect to a specified test clock cycle. That is, a signal line is associated with a *slow-to-rise* transition delay fault and a *slow-to-fall* transition delay fault.

Path delay fault model: A path delay fault assumes a transition along a path cannot propagate fast enough with respect to a specified test clock cycle.

The stuck-at and bridging fault models model the behaviors of structural defects that cause static logic value changes. With shrinking supply voltages and decreasing process feature sizes, timing-related defects that only affect the speed of a transition are on the rise. Such defects need to be modeled by the transition and path delay fault models.

9.2.1.2 Automatic Test Pattern Generation (ATPG)

Test generation is the process of generating test patterns to *detect* faults assumed under a selected fault model. As illustrated in Figure 9.1, the test pattern p for the stuck-at fault f in a circuit means that the faulty circuit with f and the fault-free circuit without f produce different logic values on at least one output. *Fault detection* in the context of other fault models can be defined in a similar manner.

Figure 9.2 shows an illustrative flow of test generation. The first step is to select a proper fault model, and all faults under the fault model are listed as target faults. In each test generation run, a *primary* fault is targeted and logic values necessary for its detection are determined by using such algorithms as D, PODEM, and FAN [7–9]. Note that not all inputs need to have logic values in order to detect a primary fault. If many inputs remain unspecified (indicated by *X-bits* in Figure 9.2), one or more *secondary* faults can be further targeted for fault detection. This process, called *dynamic compaction*, helps reduce final test pattern count. Even after dynamic compaction, some inputs may still remain unspecified. Such an input combination is called a *test cube*. *X-filling* is then used to assign logic values to X-bits in a test cube so as to create a fully specified *test pattern*. For example, *random-fill* can be conducted by assigning random logic values to all X-bits in a test cube. Random-fill has strong fortuitous fault detection capability that helps reduce final test pattern count.

The detection of a stuck-at or bridging fault needs one test pattern, while the detection of a transition or path delay fault needs a pair of test patterns. This is because in order to detect a delay fault, a path corresponding to the fault needs to be sensitized, a transition needs to be created at the start point of the path, and the test response needs to be measured at the endpoint of the path at the required timing.

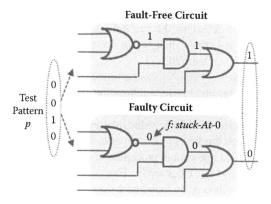

FIGURE 9.1   Concept of fault detection.

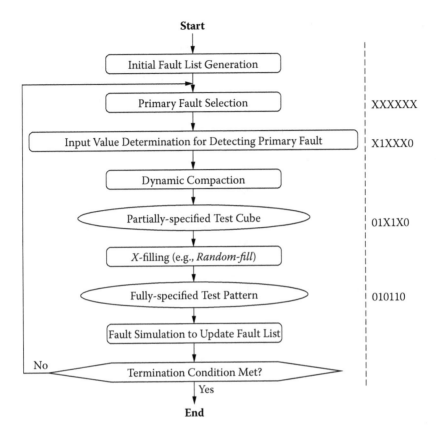

FIGURE 9.2   Illustrative flow of test generation.

9.2.1.3   Test Quality Assessment

In test generation, the quality of generated test patterns needs to be properly assessed. The most commonly used metric for this purpose is *fault coverage*, which is the ratio of faults detected by a set of test patterns to the total number of faults listed under a fault model [7–9]. Since a fault model is an indirect representation of the behaviors of physical defects, in some cases there is a need to assess the capability of test patterns in detecting unmolded physical defects. A metric for this purpose is *bridging coverage estimate* (BCE) [13], which takes into consideration the number of times that a stuck-at fault is detected by a test pattern set so as to assess the capability of the test pattern set to detect bridging-type defects. The delay test quality of a transition delay test set can be assessed by the *statistical delay quality level* (SDQL) metric [14], which takes the sizes of delay defects and the lengths of sensitized paths into consideration. SDQL is important since small-delay defects are becoming dominant timing-related defects, and transition fault coverage alone cannot accurately assess the capability of a transition delay test pattern set in detecting small-delay defects. In addition, *output deviation* can be used as an effective surrogate metric [15].

9.2.2   Basics of Design for Testability

Test generation for a large combinational circuit may take days, if not weeks, to complete. For a sequential circuit, even a modest one, its test generation time can easily become prohibitively long. This means that industrial circuits, which are mostly sequential in nature with tens of thousands of flip-flops (FFs), cannot be practically tested in their original form. A typical solution is to convert a sequential circuit into a *scan design* so that its testing can be conducted through targeting the combinational portion of the original circuit. Scan design is the most fundamental type of *design for testability* (DFT) [7–9].

In addition to making a sequential circuit easily testable, DFT also manifests itself as *built-in self-test* (BIST) [7–9], in which all test tasks (test generation, test application, and test response analysis) are conducted in the CUT by specially added test circuitry. Furthermore, DFT in the form of on-chip test stimulus decompressors and test response compressors/compactors can help reduce final test data volume dramatically in *test compression* [9–16].

In the following, the most fundamental DFT technology, i.e., scan design, and its testing (*scan testing*) are briefly described.

9.2.2.1   Scan Design

Figure 9.3(a) shows a sample sequential circuit, which consists of a combinational portion and three FFs. For the combinational portion, external inputs are called *primary inputs* (PIs) and internal inputs from the FFs are called *pseudo-primary inputs* (PPIs). In addition, external outputs are called *primary outputs* (POs) and internal outputs to the FFs are called *pseudo-primary outputs* (PPOs). Faults in the combinational portion are usually hard to detect since *PPIs* are hard to control and *PPOs* are hard to observe.

The most common DFT technique to solve this problem is scan design [7–9]. The basic idea is to replace all functional FFs with *scan FFs* and connect them into *scan chains*.

There are many different scan FF designs, and Figure 9.3(b) shows a typical one called the *MUX-D scan FF*. Functionally, a MUX is added to the input of a D-FF, resulting in two inputs to the scan FF: one being the original D input and the other being a new input called *scan input* (SI). The selection signal for the MUX, called *scan enable* (SE), determines which input is to be used to update the output Q.

In a scan design, functional FFs are first replaced with scan FFs. Then, scan FFs are connected into scan chains. An example is shown in Figure 9.3(c). Here, the D inputs of all scan FFs come from the combinational portion, while the *SI* input of each scan FF (except the first one) is connected to the Q output of the preceding scan FF in a scan chain. The *SI* input of the first scan FF in a scan chain is directly controllable from the outside, and it is called the scan input (SI) of the scan chain. On the other hand, the Q input of the last scan FF in a scan chain is directly observable from the outside, and it is called the *scan output* (SO) of the scan chain. Note that a scan design may have multiple scan chains, and its number is usually limited by the number of pins available for corresponding *SIs* and *SOs* of the scan chains. The *SE* inputs of all scan FFs in a scan chain are connected together to become the scan enable (SE) of the scan chain.

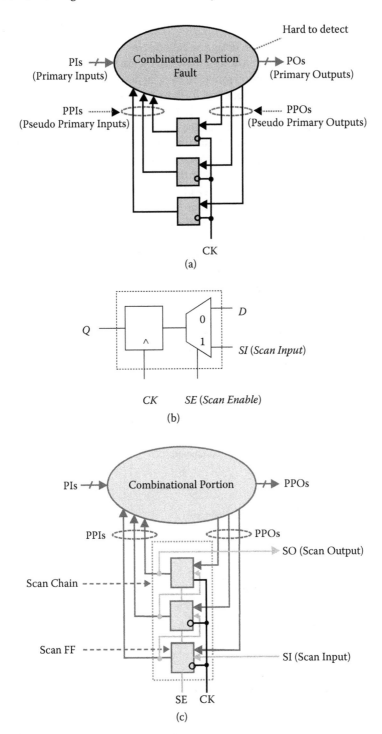

FIGURE 9.3 Example of scan design. (a) Original design. (b) Scan FF. (c) Scan design.

9.2.2.2 Scan Testing

LSI testing conducted for a scan design is called *scan testing*. Figure 9.4 illustrates scan testing for the example scan design shown in Figure 9.3(c). Basically, two operation modes (i.e., *shift* and *capture*) provided by scan design are used for scan testing. These two modes are switched back and forth by using the *SE* signal [7–9].

Shift mode is entered by setting *SE* = 1. The equivalent circuit for shift mode is shown in Figure 9.4(a). It can be seen that the scan chain operates as a shift register in shift mode, in which the *PPI* values (test stimulus) for a new test pattern are shifted in serially from *SI* and the *PPO* values (test response) for the previous test pattern are shifted out serially from *SO*. It is clear that with this shift register function provided in shift mode, test stimuli can be easily set to all *PPI*s and test responses from all *PPO*s can be easily observed. Note that *N* shift clock pulses need to be applied, where *N* is the length of the longest scan chain in the scan design.

Capture mode is entered by setting *SE* = 0. The equivalent circuit for capture mode is shown in Figure 9.4(b). Obviously, this circuit configuration is the same as its original (or functional) configuration. Basically, one or two capture clock pulses are applied to load the *PPO* values (internal part of the test response for the previous test pattern) into all FFs, in order to be shifted out in the next shift operation.

Figure 9.4(c) shows the timing waveform for one test pattern for the scan design shown in Figure 9.4(c). Three shift clock pulses are applied in shift mode since the scan chain has three scan FFs. In capture mode, one capture clock pulse is applied long after the last shift clock pulse is applied. Such scan testing is called *slow-speed*

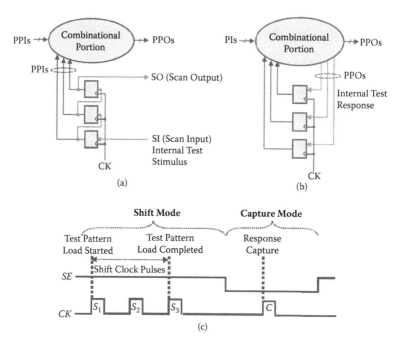

FIGURE 9.4 Example of scan testing. (a) Shift mode (*SE* = 1). (b) Capture mode (*SE* = 0). (c) Timing waveform.

scan testing, which is used to test for non-delay-type faults with test patterns generated for stuck-at faults, bridging faults, etc.

Scan testing is conducted by repeating shift and capture for all test patterns, referred to as *scan test patterns*. Note that scan test patterns are generated for the combinational portion of a sequential circuit since scan design makes it possible to efficiently detect faults in the combinational portion in a sequential circuit. Since test generation and test application for a sequential circuit are virtually impossible, scan design is truly indispensable for LSI testing.

9.2.2.3 At-Speed Scan Testing

Shrinking feature sizes and increasing clock frequencies have made timing-related defects a major cause for failing LSI circuits. Testing for such defects needs delay test patterns, which are usually generated with the transition delay fault model or the path delay fault model. In addition to delay test patterns, delay test application is usually conducted in the form of at-speed scan testing, in which a transition is created at the start point of a path and whether it can propagate to the endpoint of the path in a functional clock cycle is checked by making use of scan design [9].

There are two schemes available for implementing at-speed scan testing, *launch-off-shift* (LOS) and *launch-off-capture* (LOC), as described below.

The LOS scheme is illustrated in Figure 9.5(a). This scheme generates transitions at the start points of sensitized paths by the differences of logic values loaded into FFs by

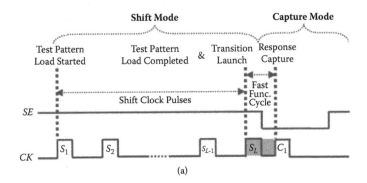

(a)

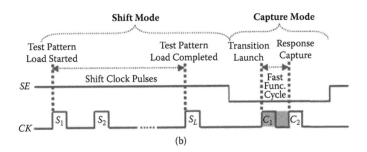

(b)

FIGURE 9.5 Clocking schemes for at-speed scan testing. (a) Launch-off-shift (LOS) scheme. (b) Launch-off-capture (LOC) scheme.

the next-to-last shift clock pulse, S_{L-1}, and the last shift clock pulse, S_L. Test responses are captured by the capture clock pulse, C_1. Note that the time between the last shift clock pulse, S_L, and the capture clock pulse, C_1, is set to be equal to the functional cycle in order to realize at-speed scan testing. Since only one capture clock pulse is used in capture mode, the LOS scheme is also referred to as the *single-capture* scheme.

The LOC scheme is illustrated in Figure 9.5(b). Different from the LOS scheme, this scheme generates transitions at the start points of sensitized paths by the differences of logic values loaded into FFs by the last shift clock pulse, S_L, and the first capture clock pulse, C_1. Test responses are captured by the second capture clock pulse, C_2. Note that the time between the first capture clock pulse, C_1, and the second capture clock pulse, C_2, is set to be equal to the functional cycle in order to realize at-speed scan testing. Since two capture clock pulses are used in capture mode, the LOC scheme is also referred to as the *double-capture* scheme.

Generally, the LOC scheme is easier to implement than the LOS scheme since the latter requires an at-speed *SE* signal that needs to be treated in a similar way as a clock signal in physical design. On the other hand, the LOS scheme usually achieves higher fault coverage than the LOC scheme. This is because it provides better controllability and observability due to the use of a single-capture clock.

Transition delay test patterns are widely used for at-speed scan testing. However, high transition fault coverage may not directly translate into high at-speed scan test quality. This is because at-speed scan test quality also depends on the relations among (1) the lengths of sensitized paths, (2) the extra delays caused by timing-related defects, and (3) the length of the test cycle, as illustrated in Figure 9.6. Especially, shrinking process feature sizes and increasing clock frequencies have increased the occurrence possibility of small-delay defects, which only cause a slightly increased delay at the defect site [17]. As shown in Figure 9.6, if the sensitized path passing through a small-delay defect site is relatively short with respect to the test cycle, at-speed scan testing cannot detect such a defect. Only when the sensitized path is long enough can such a small-delay defect be detected in at-speed scan testing. In order to achieve this goal, transition test generation needs to make sensitized paths as long as possible.

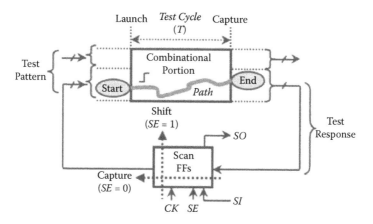

FIGURE 9.6   Sensitized path and at-speed scan test quality.

Such test generation is called *small-delay test generation* [18]. The quality of a small-delay test pattern set can be assessed with the SDQL metric [14].

9.3 TEST POWER ISSUES

As integrated circuit feature size continues to shrink and the number of wireless and portable devices grow, power consumption has become a critical element for the semiconductor industry to address during both functional operation and manufacturing test.

During the functional operation, high power consumption implies:

- Higher design and manufacturing cost due to the extra effort to calibrate a power grid to meet the power supply requirement
- Higher system cost due to the packaging and cooling requirements
- Shorter device life cycle and lower device reliability
- Shorter battery life for portable devices

In order to reduce the functional power consumption, various low-power management techniques have to be implemented during the design phase [19–23]:

Power domain: The device is implemented by including multiple functional blocks. Each functional block can be independently powered up or down through controlling the power switches used to gate the power supply connection to each functional block.

Multiple supply voltages: Depending on operation conditions, a power domain can be operated under the different supply voltages. The power domains operated at different supply voltage levels are connected through level shifters.

Isolation logic: It is inserted at the boundary connecting two power domains in order to isolate the power-off domain from the power-on domain.

State retention cell: A special sequential state element that preserves its state when the power domain containing this sequential state element is powered down.

Power mode: It is a functional operation mode that makes each power domain operate at either the power-on or the power-off state and assigns specific supply voltages to each power-on domain.

Clock gating: A powerful technique to reduce the power consumption in a power-on domain through dynamically blocking the clock pulse to reach a set of sequential state elements. The clock gating is often implemented hierarchically in order to improve the flexibility to control the power consumption during functional operation.

To meet the design methodology changes in the low-power designs, traditional DFT methodologies have to become power-aware such that:

- The power constraints should not be violated by the test programs and the DFT architecture inserted into the design.
- Power management circuitry, including isolation logic, state retention cells, level shifters, power switches, etc., has to be tested adequately.

- ATPG should effectively utilize the embedded clock gating logic to reduce the test power.
- DFT hardware added to generate the low-power test patterns and improve the testability of the low-power management circuitry should minimize its area overhead and avoid its impact on system performance while maximizing the benefits to reduce the test power and the test cost.
- A test scheduling plan should cover all power domains, glue logic, and all supply voltage levels while minimizing test application time and test data volume and reducing overall test power consumption.

Besides taking functional power management into account, it is also necessary to make the test power consumption close to the worst-case functional power consumption in order to avoid yield loss, reliability degradation, and permanent damage of a circuit. It has been shown that scan-based tests significantly reduce the test cost and achieve satisfying test quality compared to functional tests. However, it was also observed that the scan-based tests may cause circuit switching activity in excess of the activity during normal operation. The main reason is that the scan test patterns often make the circuit operate in nonfunctional states. On the other hand, as the small-delay defects become one of the dominant defect types introduced during manufacture due to the process variation, an at-speed test set based on the transition fault model becomes mandatory to be included in the test suites in order to achieve adequate test quality. It has been shown that power supply droop issues caused by the excessive test power consumption reduce the effectiveness of at-speed scan testing, resulting in test escape due to clock stretch [24] and yield loss due to incorrect capture caused by additional gate delay [25]. Test power reduction in the scan-based tests has become more urgent in today's nanometer designs.

In the scan-based tests, there are two types of test power: average test power and peak test power. The former one is the ratio of consumed energy over a time period to the test time, while the latter one is the highest power value at any given instance [5].

Excessive average test power consumption may cause the problems listed below [5]:

- The increased thermal load on the circuit under test is likely to create hot spots that damage the silicon, the bonding wires, or even the package.
- Elevated temperature and current density in test mode will increase electromigration, which in turn causes more intensive erosion of conductors that severely decrease the reliability of the device.
- It is imperative to use an expensive package to tolerate excessive heat during test.

When the peak power is beyond the design's limit, the circuit cannot be guaranteed to function correctly since the supply voltage droops introduce additional gate delay and may cause false failure.

The test power consumed during scan shift and during capture is referred to as shift power and capture power, respectively. The average power consumption is determined by the shift power. Excessive shift power accumulation may cause scan chain failure and impact the test responses captured in the capture cycles. During capture, the power reduction is typically focused on the peak power since it may cause supply voltage droops resulting in yield loss or test escape due to the additional gate delays or the clock stretch.

9.4 POWER SUPPLY NOISE

Along with the integrated circuit feature size scaling, the supply voltage between the V_{dd} and V_{ss} pins of a standard cell is decreased in order to reduce functional power consumption. It leads to the standard cell having a lower margin to tolerate the power supply noise. Since the supply voltage level is one of the main factors to determine the delays of each cell in the design, the extra delay introduced by the supply noise may causes chips to fail during functional operation. This phenomenon would be even more problematic during structural-based at-speed testing, as the testing power is often higher than the functional power, and it could result in yield loss due to falsely identifying good chips as faulty.

In this section, we present an overview of the impacts of the power supply noise on at-speed delay tests, including the phenomenon of power supply droop, various on-chip instruments to measure the power supply droop, the impacts of the power supply droop on at-speed scan testing, and test power estimation for scan test patterns.

9.4.1 STRUCTURAL AT-SPEED DELAY TESTING

For deep-submicron technology, test sets generated based on traditional static fault models, including stuck-at fault and I_{ddq}, are not sufficient to maintain the test quality. At-speed delay testing has become mandatory in industry to detect delay defects that cause timing violations during functional operation.

Path delay fault and transition fault are two common fault models used by the ATPG tools to generate scan test patterns detecting the delay defects. The path delay fault model targets cumulative delays along specific paths in the design, while the transition fault model targets excessive delay on individual gates or nets. To detect the delay defects, a scan test pattern includes at least two at-speed clock cycles that create a transition at the target gate and propagate the fault effect to an observation point.

Depending on how the transition at the target gate terminal is launched, there are two types of test schemes, launch-off-shift (LOS) and launch-off-capture (LOC). In the LOS, the transition is launched in the last shift cycle during scan loading, and it requires the scan enable signal to be changed at functional frequency. In the LOC, the first capture clock pulse is used to launch the transition, and it relaxes the requirement about changing the scan enable signal at-speed.

In order to reduce test cost, it is desirable to utilize a low-speed tester to test high-frequency devices at-speed. This goal is achieved by utilizing on-chip phase-locked loop (PLL) to generate at-speed test clocks during capture. Since the clock sources have to be switched from external low-speed shift clocks to internal at-speed PLL clocks, dead cycles are often needed between the last shift cycle and the first capture cycle in order to synchronize the clock sources. It makes the LOC the preferred at-speed test scheme in industry today.

9.4.2 POWER SUPPLY DROOP

In CMOS design, power is distributed to the circuit nodes through power grids. The parasitic components embedded in the power grids, including resistance, capacitance, and inductance, cause power supply noise when the circuit nodes switch their states

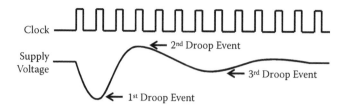

FIGURE 9.7   Supply voltage droop in response to active clock.

in response to active clocks that draw current from power supply to power ground. A simplified power supply noise model includes two components, resistive noise, $Ri(t)$ (also known as IR-drop), and inductance noise, $L(di/dt)$.

The resistance noise depends on the amount of current drawn from the power supply, and it shows up as power grid voltage decrease due to the resistances between the power supply and the circuit nodes.

The inductance noise is caused by the instantaneous change in current drawn from the power supply [26], which produces resonant voltage spikes, i.e., voltage droop events, as shown in Figure 9.7. The voltage spikes tend to have fast edges and are associated with very-high-frequency components. They will die away over time. When the switching activity causes the current to change faster than the RLC time constant and the circuit is operated in high frequency, it will take multiple clock cycles to recover the supply voltage level to be within the desired range.

As shown in Figure 9.7, the first voltage droop event typically has a resonant frequency between 100 MHz and 1 GHz and its frequency is determined by the characteristics of the circuit's package. The second and third droop events have resonant frequencies of 1–10 MHz and 1–100 KHz, respectively, and they are determined by the package, the motherboard levels' decoupling, and the placement of voltage regulator modules [27, 28]. In summary, the voltage droop waveform is independent of the circuit switching activity, but is fixed for a particular design. However, the magnitude of the first droop event is proportional to the clock frequency. The higher the clock frequency is, the more voltage droop the circuit will experience. Moreover, higher circuit switching activity will induce larger voltage droop.

9.4.3   ON-CHIP MEASUREMENT FOR POWER SUPPLY VOLTAGE DROOP

Quantifying the power supply drop effects can be done either based on simulation or by using on-chip instruments. Since the interaction between the power supply droop and the delays at the clock and the logic paths is very complex and the power supply droop is a transient phenomenon, simulation-based methods cannot accurately evaluate the supply voltage droop caused by the switching activity. Using on-chip instruments is the only method to monitor the transient circuit state precisely. Since the gate delay is directly determined by the power supply voltage, an on-chip instrument typically performs the delay measurement to estimate the supply voltage value. During the measurement, a test or instruction sequence that induces the power supply noise is run repetitively. The circuit state is captured in scan cells and shifted out for further analysis at the end of the measurement.

To study the clock stretch during at-speed scan testing, the calibration circuit shown in Figure 9.8 was proposed in [24]. In this circuit, a programmable combinational delay line is used to set the different delays between two DFFs. The delay measurement is performed for two sequences: one is operated in the function mode and the other is a two-cycle LOC scan pattern. Through observing the signal *Fail*, the configuration of the delay line at the pass and fail boundary is determined for every sequence. The delay difference between two boundary configurations is a measure of clock stretch.

An on-chip macro, called Skitter, was used to measure timing uncertainty on an IBM microprocessor [29]. As shown in Figure 9.9, a delay line including 129 invertors is implemented to achieve 5–8 ps resolution. When the clock traverses through the delay line, the responses of each invertor are captured by 128 sampling latches for every clock cycle. The clock edge is detected by two adjacent sampling latches with the same captured value. The number of sampling latches between two adjacent full cycle edges is a measure of clock cycle. The smaller this number is, the shorter the cycle is; i.e., the lower supply voltage caused by the voltage droop makes the clock edge travel through fewer inverters.

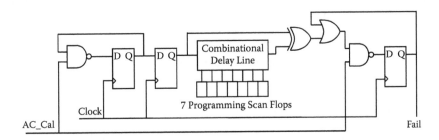

FIGURE 9.8   Delay calibration circuit. (From Rearick and Rodgers, Calibrating clock stretch during AC scan test, In *Proceedings of International Test Conference*, 2005, Paper 11.3.)

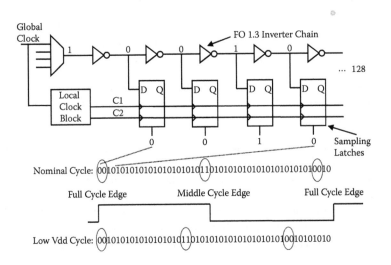

FIGURE 9.9   Skitter circuit. (From Franch et al., On-chip timing uncertainty measurements on IBM microprocessors, In *Proceedings of International Test Conference*, 2007.)

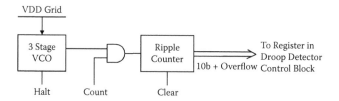

FIGURE 9.10 VCO-based droop detector circuit. (From Petersen et al., Voltage transient detection and induction for debug and test, In *Proceedings of International Test Conference*, 2009, Paper 8.2.)

In [30], the local power supply grid is used as control input to a three-stage voltage-controlled oscillator (VCO), as shown in Figure 9.10. The VCO converts the voltage at the power grid into a high-frequency digital signal with the frequency proportional to its input voltage. A ripper counter is used to count the number of cycles produced by the VCO within a time window. The data stored in the counter at the end of the measurement window reflect the power supply voltage change while applying a clock sequence, including one or multiple clock cycles. The VCO-based detector provides 10–15 mv resolution.

To monitor the supply voltage changes across the design, it often needs multiple on-chip instruments spread in different parts of the design.

9.4.4 Impacts of Power Supply Droop on At-Speed Scan Testing

During functional operation, a steady stream of clock pulses is applied and adjacent clock cycles typically have a constant level of switching activity. It makes the power supply voltage have less variation, but the average voltage level may be a little lower than the nominal value. However, an at-speed scan test pattern is applied in two stages:

Shift mode: The test stimuli are shifted into scan chains through slower clocks.
Capture mode: One or more burst clock cycles running at functional speed are applied to capture the fault effect in the scan cells.

When switching from the shift mode to the capture mode, the increased clock frequency and the latency between the last shift clock pulse and the first capture clock pulse will cause an abrupt change of the circuit state:

- The trough of the first voltage droop event shown in Figure 9.7 becomes deeper when the capture clock frequency increases.
- When the circuit stays at a state with no switching activity for some clock cycles between the last shift cycle and the first capture cycle, pausing the clocks pulls up the supply voltage. When the time interval between the last shift and the first capture is not large enough, the first capture pulse will start from a voltage higher than the nominal voltage [31]. Figure 9.11 shows this phenomenon.

In modern design, the at-speed capture clocks are typically generated by on-chip PLL so as to use a low-cost tester to test fast chips as well as improve the precision of

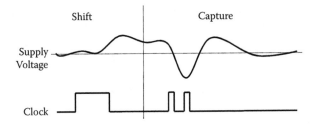

FIGURE 9.11 Pull-up effect of supply voltage.

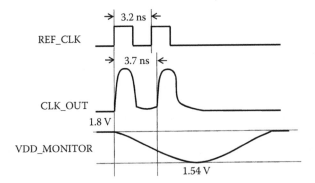

FIGURE 9.12 Clock stretch caused by supply voltage droop. (From Rearick and Rodgers, *Calibrating clock stretch during AC scan test*, In *Proceedings of International Test Conference*, 2005, Paper 11.3.)

the test clocks. It often takes some PLL clock cycles to switch the clock sources from external slow-shift clocks to internal PLL clocks. The latency induced by the clock synchronization is hard to be avoided. On the other hand, it is desirable to start the first capture clock pulse from the nominal supply voltage. Sufficient latency allows the power rails to be fully charged before capture. However, the latency exacerbates the power supply droop due to an abrupt change of the circuit state from quiet to active.

The study in [24] illustrates that the capture clock period varies by more than 15% due to supply voltage droop. As shown in Figure 9.12 given in [24], the supply voltage drops from 1.8 V to 1.54 V, and it stretches the first capture clock period by 0.5 ns during at-speed scan testing. Since the second clock pulse is delayed, it increases the chances that delay defects escape from scan testing. Consequently, the scan test patterns may pass at clock frequencies higher than the functional patterns.

Since higher switching activity may cause larger power supply droop, the impact of the switching activity on path delays was investigated in [32, 33]. By using different fill strategies to specify the don't care bits of the same test cube, the methods in [32, 33] generated multiple scan test patterns with different switching activity and applied them to test the same path delay fault. The silicon data collected from a design fabricated by 130 nm technology show a 10% voltage droop can cause up to a 15% delay increase of the path under test, and the impact of the voltage droop on longer paths is more serious than on short paths [32]. The same experiments were carried out in [33] for a 65 nm technology design by using three different supply voltages. The silicon data show an impact similar to that observed for the 130 nm design. An additional experiment carried out in [33] demonstrates that a path in the region with higher local switching activity is much slower than an almost identical path in the region with lower local switching activity, even if the global switching activity is the same. The above experimental results illustrate the scan test patterns may fail at frequencies lower than the functional patterns if they have much higher switching activity than the functional mode.

A theory was developed in [26] to characterize the effects of power supply droop on at-speed scan testing during capture. The key factor behind the theory is that the period of voltage droop events is determined by the package characteristics of a design, but the peak voltage droop is proportional to the burst clock frequency. The theory also assumes that the power supply rails are fully charged to their nominal value before capture.

Figure 9.13 shows the relative relationship between voltage droop events and clocks with different frequencies. For the clock with higher frequency, the second capture clock finishes pulsing far ahead before the first voltage droop event is fully deployed. The supply voltage level seen by the capture clock pulses may be higher than the average voltage level during the functional mode. Consequently, the scan test patterns pass at the clock frequencies higher than the functional patterns. It could lead to test escape. For the clock with lower frequency, the second capture clock starts a pulse near the peak of the first voltage droop event. Although the peak voltage droop level reduces as the clock becomes slower, it is still larger than the voltage droop seen by the faster clock. Consequently, the scan test patterns fail at the clock frequencies lower than the functional patterns. It could lead to yield loss.

When studying the impacts of supply voltage droop on at-speed scan testing, both the operating voltage level and the power delivery system play important roles. In [26], the comparison of the maximum operation frequency, f_{max}, between functional patterns and transition scan test patterns shows that the transition scan test patterns run faster than the functional patterns at higher voltage. The f_{max} gap is reduced as the operating voltage decreases. At very low voltage, the transition scan test patterns become slower than the functional patterns. On the other hand,

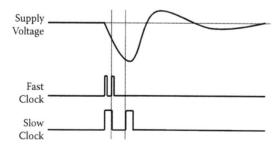

FIGURE 9.13 Voltage droop events versus clocks with different frequencies. (From Pant and Zelman, Understanding power supply droop during at-speed scan testing, In *VLSI Test Symposium*, 2009, pp. 227–232.)

the robustness of the power delivery system also determines the developing time and the recovering time of the supply voltage droop. When the circuit experiences a sudden change of activity, a power network with poor current delivery capacity will suffer from a faster first-voltage droop event, and it will take longer to recover [34]. Consequently, the at-speed scan patterns are more prone to fail at a clock frequency lower than the functional mode when they have higher switching activity.

9.4.5 TEST POWER ESTIMATION

To quantify the power dissipation consumed by test patterns, several test power estimation methods have been proposed. They can be classified into two categories:

Non-timing-based approach: Non-timing-based logic simulation is used in this approach. The switching activity occurring at a gate in a clock cycle is weighted based on the gate load. The sum of the weighed switching activity (WSA) among all the gates is used as the test power estimation for the clock cycle. Typically, the weight can be chosen as a unit weight, the number of a gate's fan-out plus 1, or the load capacitance.

Timing-based approach: In each clock cycle, a timing-based simulation is used to record a specific time instance that the gate switches. The clock cycle is divided into multiple time slices, and the WSA is calculated for each time slice. The time slice with the maximum WSA is used as the peak test power estimation in this clock cycle. Since the temporary distribution of the transitions is more important to evaluate the impacts of the IR-drop, the timing-based approach is more accurate than the non-timing-based approach. However, it suffers from higher computational complexity. On the other hand, without considering the physical layout, both approaches only provide the quantitative relation between global switching activity and IR-drop. They cannot distinguish the IR-drop impacts of two test patterns with the same level of global switching activity. In fact, the test pattern with higher regional switching activity in timing critical paths is more likely to suffer from IR-drop than the test pattern with evenly distributed switching activity.

To evaluate IR-drop effectively, techniques such as the one proposed in [35] divide design cells in a chip into groups according to the chip's physical layout. The regional switching activity provides more insight about IR-drop severity at hot spot regions. However, grouping design cells based on physical layout alone may not provide adequate spatial resolution to expose the real regional IR-drop problem. The realistic power grid structure has to be taken into account as well since the higher the number of cells driven by the same power supply pad that have switching, the more severely the IR-drop will impact the performance of those cells. Combining the power grid structure and the cells grouped based on the power supply pads gives a more accurate power estimation model to reveal the IR-drop issue. A weight assignment scheme based on this model was proposed in [36]. To speed up the IR-drop calculation in [36], a test pattern-independent characterization phase is carried out for each design cell in order to calculate the cell weight vector with respect to every power branch in the power grid under the assumption that only the cell under consideration has switching activity. The IR-drop profile for a test pattern on the power grid is calculated by superpositioning the weight vectors of all the cells with switching.

When guiding ATPG to generate low-power test patterns, it requires that the test power estimation approach not only has less impact on ATPG performance, but also provides meaningful accuracy. As a result, the non-timing-based approach is often preferred.

To avoid time-consuming simulation of each shift cycle, the transitions occurring at each scan cell are weighted to estimate the shift power by ATPG tools. The test stimulus in the current test pattern and the test response captured from the previous test pattern are considered simultaneously in order to report the shift power more accurately. For a scan chain j including C scan cells, its weighted transition is calculated as follows:

$$WT_j = \sum_{i=2}^{C} (r_i \oplus r_{i-1}) \cdot (C - i + 1) + (t_s \oplus r_1) \cdot C$$

$$+ \sum_{i=C+1}^{S} (t_i \oplus t_{i-1}) \cdot C + \sum_{i=2}^{C} (t_i \oplus t_{i-1}) \cdot (i - 1) \tag{9.1}$$

where S is the number of shift cycles and $S \geq C$; $r_i, i \in \{1, ..., C\}$, is the test response captured at scan cell i after applying the previous test pattern; $t_i, i \in \{1, ..., C\}$, is the test stimulus assigned to scan cell i at the end of the shift operation; and $t_i, i \in \{(C + 1), ..., S\}$, is the test stimulus shifting into the scan chain when the number of scan cells in the scan chain is less than the number of shift cycles. Moreover, scan cell 1 in a scan chain is the cell driven by the scan input pin. The shift power is equal to the sum of the weighted transitions among all the scan chains.

To estimate the capture power, the ATPG tools carry out non-timing-based logic simulation to determine the switching activity of each gate in each clock cycle. A weight equal to the number of gate fan-outs plus 1 is typically used to calculate the WSA.

9.5   TESTING OF POWER MANAGEMENT CIRCUITRY

In low-power designs, additional circuitry is included to manage the power supply for different functional blocks during functional operation. The key components included in a typical power management circuitry are shown in Figure 9.14, and their functions were described in Section 9.3.

The structure and the operation of the power management circuitry can be described by either Unified Power Format (UPF), which is going to become IEEE Standard P1801, or Common Power Format (CPF), which was developed by Cadence. The main contents defined in the UPF/CPU include:

- A list of power domains and the control method for their power switching logic
- A list of power modes and the operations and the supply voltages for each power domain under each power mode

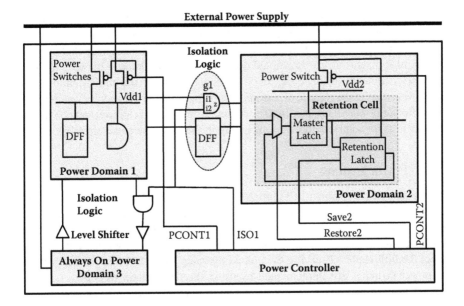

FIGURE 9.14   Functional power management circuitry.

- Logic connection among power domains through isolation logic or level shifters
- A list of retention cells included in each power domain and the method to save (restore) the state of the retention cells when the power domain is powered down (up)

The power management circuitry has to be considered during testing. Before applying DFT tools to generate the test patterns for it, the DFT tools have to check a set of rules in order to ensure the scan-based tests are effective and applicable for the design with power management circuitry. Typically, these rules do the following checks:

- The power data loaded from UPF/CPF are correct and cover the operations of every power domain under every power mode.
- The scan chain spanning multiple power domains cannot power down some of its scan cells and power up the rest simultaneously.
- All control logic for the scan operation must be powered on during scan shift.
- The power domains under test cannot be incorrectly powered down during capture.
- All power control signals, including isolation signals, retention save/restore signals, power switching control signals, etc., must be generated from always-on power domain(s).
- A level shifter must be inserted at the paths connecting the power domains operating in different supply voltages.
- An isolation cell must be inserted at the paths connecting the power domains that can be powered down independently.

To improve the testability for the power management circuitry, a DFT structure such as the power test access mechanism (PTAM) [20] can be inserted in the design by using a DFT insertion tool. During test, this structure generates signals that override the control signals from the functional power controller in order to improve the flexibility to schedule the testing order for power domains, ensure the power domains hold their power state in the middle of testing, ensure the isolation between power domains with different power states, etc. In this section, we provide a brief overview on how to generate tests for some components, including in the power manager circuitry.

9.5.1 Power Switches

The power switches shown in Figure 9.14 are used to control the power state of a power domain. Testing the power switches can be performed either implicitly or directly. The implicit test is done when the power domain gets tested. For example, to test a power switch stuck open, one can use the test patterns for retention cells and detect the failures by observing the unloading value from the retention cells. When the power switches are implemented by using the daisy chain as shown in power domain 1 in Figure 9.14, delay testing can be used to test part of the power switches stuck open. This strategy does not require additional DFT hardware, but it has difficulty diagnosing the faulty switches implemented by using the daisy chain. The preferred method is to insert DFT control and observation points at the power switches in order to deterministically detect failures through comparison of the voltage at each power switch output with a reference voltage. One of the DFT implementations can be found in [37]. However, this technique suffers from long discharge time that may lead to either false pass or long test time. In [38], the extra discharge transistors are added to reduce the discharge time during test application.

Although the power switches are effective in reducing the static power by powering off the logic blocks in idle mode, it requires a long wake-up time when reactivating the logic blocks from the idle mode. The intermediate strength power gating was proposed in [39] to achieve the trade-off between suppression of leakage current and shorter wake-up time. To test this type of power switching structure, a signature analysis technique was proposed in [40]. The proposed DFT technique converts the voltage at virtual power rail or virtual ground rail into a frequency reading. The converted data are shifted out from the scan chain in order to detect and diagnose the failure power switches.

9.5.2 Isolation Cells

Testing the isolation logic located between two power domains completely requires:

* A power mode that can power up both power domains
* An isolation enable pin that can be controlled by the test generator directly

For example, to test the isolation AND gate g_1 shown in Figure 9.14, the functional power controller should be configured to the power mode that powers up power domains 1 and 2, and to set the isolation, enable signal $ISO1$ to 1. Under this configuration, the test generator generates tests for the faults at g_1/i_1 and g_1/z and the fault g_1/i_2 s-a-0.

To detect the fault g_1/i_2 s-a-1, the functional power controller has to set *ISO*1 to 0, and it requires a power mode that powers up power domain 2 and powers down power domain 1. Due to the unknown value at g_1/i_1, the fault is untestable. To generate a test deterministically for this fault, a DFT structure such as PTAM is needed to enable the test generator to take control of *ISO*1 when both power domains are powered up.

9.5.3 RETENTION CELLS

To test the retention cells included in a power domain, a power mode that powers up this power domain is selected first in order to ensure the scan shift operation can load and unload known value from the retention cells. During capture, a test sequence includes the following subsequences:

- Applying a power-down sequence to take the power domain to an off state. The power-down sequence issues a control signal to save the value loaded into the retention cell. For example, to test the retention cell shown in Figure 9.14, the control signal *Save2* is activated to transfer the value from the scanned master latch to the retention latch before powering down the power domain.
- Keeping the power domain in an off state for some clock cycles.
- Applying a power-up sequence to take the power domain back to an on state. The power-up sequence issues a control signal to restore the value saved in the retention cell. For example, the control signal *Restore2* in Figure 9.14 is activated to restore the value from the retention latch to the master latch after powering up the power domain.

In commercial ATPG tools, the capture test sequence can be defined in the test procedure and the ATPG tool will automatically generate retention test patterns based on it in order to verify the capability of the retention cells retaining 0 and 1. In the commercial ATPG tool *FastScan®*, the capture test sequence for retention test is described by using a named capture procedure that defines the complete clock sequence between scan loading and unloading.

9.5.4 LEVEL SHIFTERS

The level shifter is typically modeled as a buffer for test generation purposes. To generate tests for a level shifter located between two power domains, it requires a power mode that powers up both power domains. Moreover, if the power domain can operate in multiple supply voltages, the level shifter needs to be tested repeatedly under different supply voltages.

9.6 LOW-POWER TEST SOLUTIONS

A lot of DFT techniques had been proposed in the past to reduce the test power, and most of these techniques require additional test cost. When choosing the low-power test techniques for a design, one has to consider the trade-off among test application time, test data volume, area overhead, performance impact, etc.

Typically, the low-power DFT techniques used in industry and supported by the commercial tools are based on X-fill strategies, clock gating control, blocking scan cell outputs, and clock application schemes. These techniques can be applied alone or together. We describe some of them in this section.

9.6.1 X-FILL STRATEGIES

X-fill utilizes the unspecified bits in a test cube to reduce shift power, capture power, or both simultaneously. The main advantage of using X-fill strategies is that they are a nonintrusive method, have no impact on system performance, and require no additional hardware if applying in a noncompression environment.

The motivation of using X-fill strategies comes from the observation that a test cube typically contains less than 1% of care bits. The unspecified bits in the test cube can be filled with any values without losing test coverage. However, comparing to the random-fill that fills the unspecified bits randomly, the X-fill typically generates more test patterns due to the lower probability of detecting faults by chance.

To reduce the shift power, 0-fill and 1-fill assign 0 and 1 to X bits, respectively, while adjacent fill assigns X-bits to be the same as the nearest care bit [41].

When doing test generation in the test compression environment, X-fill is less effective due to the limitation of encoding capacity enforced by the on-chip decompressor. Without additional DFT hardware, it is even inapplicable when the compression ratio is high. In [42], the low shift power decompressor was proposed by inserting a low shift power controller between decompressor outputs and scan chain inputs. Based on the observation that the majority of scan chains have no specified bit in each generated test cube, this method fills a large percentage of scan chains with constant 0. Figure 9.15 shows the low shift power controller that approximately

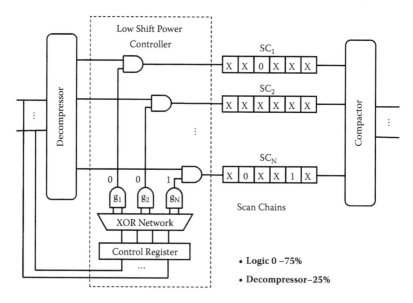

FIGURE 9.15   Constant 0-fill in compression environment.

fills 75% of scan chains with constant 0 during scan shift. The logic values generated at the outputs of gates g_1 to g_N are dynamically determined by the generated test cube. For example, the test cube shown in Figure 9.15 specifies one scan cell with loading value other than 0 in the scan chain SC_N. Therefore, the low shift power controller generates 1 at the output of g_N such that SC_N is driven by the decompressor. In the scan chains SC_1 and SC_2, the scan cells have either a unspecified value or a loading value of 0. Enforcing $g_1 = 0$ and $g_2 = 0$ makes the constant 0 shift into these two scan chains such that the shift power is reduced.

In order to support the adjacent fill in the test compression environment, the method proposed in [43] inserts a shadow register between the decompressor outputs and the scan chain inputs such that the test stimuli feeding the scan chains could be held for some shift cycles during scan shift.

When considering the capture power reduction, the X-fill techniques try to reduce the hamming distance of the logic values captured in the sequential state elements between adjacent clock cycles. The less the number of sequential state elements change their state during capture, the lower the switching activity is that occurs in the combinational logic.

In low capture fill (LCP) [44], 0s and 1s are assigned to the X-bits in the test cube iteratively by matching current and next states of the state elements in order to reduce the number of transitions at the outputs of state elements. Due to the iterative nature of this method, the runtimes to fully specified test cubes increase with the size of the circuit. Instead, the preferred fill [45] assigns known values to all unspecified bits in the test cube in a single step, and the filled values assigned to each scan cell are derived from the signal probability measurement by using a test cube-independent method. Both the LCP fill and the preferred fill achieve a similar reduction in switching activity during capture.

The aforementioned X-fill methods try to achieve global capture power reduction. However, the IR-drop-induced extra delay most likely causes the critical paths to malfunction. To address this issue, the X-fill method proposed in [46] focuses on reducing the switching activity at the gates close to the critical paths, called critical gates. Similar to the LCP fill, the X-bits are filled incrementally. In each iteration, one X-bit and its desired filling value that achieve maximal reduction on the switching activity at the critical gates are determined and assigned. The switching activity reduction metric is derived from the distance between the critical gates and the critical paths, as well as the probability of activating the critical paths. This method does not take the power grid topology into account. The distance between the logic gates is calculated based on logic level. As a result, the impact of the critical gates on the critical paths may not be estimated accurately since the critical gates and critical paths may be powered by different power supply strips or rails.

In [47], the design is partitioned into several regions according to its physical layout based on the power grid topology. Each region is bounded by power strips and limited power rails. The cells within the same region are powered by the same strips and rails. The X-bits in the test cubes are specified to minimize the switching activity within each region. Similar to the method in [46], the X-bits and their values are specified iteratively. During X-bit selection, a subset of X-bits impacting the region with a serious IR-drop effect is chosen in each iteration, and their filled values are calculated by solving pseudo-Boolean constraints in order to avoid assigning incorrect values at the target gates. The experimental results show that this method achieves better average-weighted switching activity reduction than [46], but it requires longer running time.

Due to the limitation of the encoding capacity, it may be inapplicable to reduce the capture power based on the aforementioned techniques in the test compression environment when the compression ratio is high. Fortunately, it has been observed that filling 0 for majority scan chains can reduce the capture power significantly. Therefore, the low shift power controller shown in Figure 9.15 reduces the capture power as a by-product. Moreover, the experimental results also show it helps to reduce the switching activity caused by the test responses when shifting them out from the scan chains.

9.6.2 Clock Gating Control

To control the power consumption in the power-on domains during functional operation, the low-power designs often insert clock gaters in the clock paths during logic synthesis to dynamically block the clock pulse reach to sequential state elements. During test generation, the ATPG tool can utilize these existing clock gaters to reduce the capture power consumption.

Figure 9.16 shows a typical clock gater architecture embedded in a design. During shift, *TEST_EN* is asserted that makes all clock gaters transparent in order to allow *CLK* to pass through the clock gater to shift the scan chains. During capture, *TEST_EN* is deasserted and the clock gater is controlled by its functional enable pin *FUNC_EN*. To reduce the capture power, the test generator can deassert the *FUNC_EN* if the sequential state elements controlled by the clock gater are not used to activate, propagate, or observe targeted faults. In order to effectively control the clock gaters during test generation, the sequential state elements controlling *FUNC_EN* should be scan cells. To maximize the flexibility of the power control during functional operation, clock gaters are often designed in a hierarchical way in order to achieve both fine and coarse levels of control granularity, as shown in Figure 9.17.

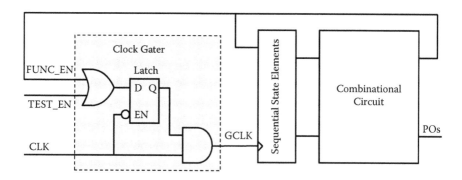

FIGURE 9.16   Clock gater architecture.

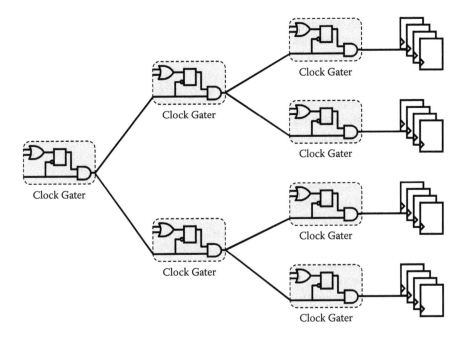

FIGURE 9.17 Hierarchical clock gating architecture.

This kind of architecture is preferred, especially when doing the test generation in a test compression environment. With the coarse control, fewer control bits are needed to prevent the clock pulse from reaching to a large percentage of sequential state elements. As a result, less care bits are needed to be encoded by the decompressor. With the fine control, the test generator can utilize spare encoding capacity to reduce the capture power further. On the other hand, the control conditions for the clock gaters at a low level are not always unique or mutually exclusive. It is possible that one condition deasserting a clock gater asserts another clock gater, although the condition deasserting both clock gaters exists. Without a higher-level clock gater that controls these two clock gaters simultaneously, it is hard for the test generator to always find the solution that deasserts both of them when they are not used to detect the targeted faults. The hierarchical clock gaters make the test generator more effective to generate test patterns with lower capture power.

The test generation procedures that utilize the clock gaters to reduce the capture power can be found in [42, 48]. The main steps included in those procedures are summarized as follows:

- Automatically identify the clock gating logic.
- Before doing test generation, the preprocessing step generates a set of test cubes that deassert the *FUNC_EN* pins controlling one or more clock gaters.
- During test generation, the test cube detecting the targeted faults is merged with the test cubes that disable as many clock gaters as possible.

Besides reducing the capture power during test generation, the set of test cubes for controlling the clock gaters can also be used for the purposes listed below:

- Providing a fast but approximate way to estimate the power consumption for a partially specified test cube such that the generated test cube detecting a set of target faults can meet a predefined switching threshold.
- Estimating the effectiveness of utilizing clock gaters to reduce the capture power. This is achieved by counting the percentage of the sequential state elements controlled by the clock gaters as well as the granularity of the number of the sequential state elements controlled by each test cube.
- Estimating the impact on test compression by counting the number of care bits included in each clock gater control test cube.

Similar to X-fill, utilizing the clock gaters to reduce the capture power often results in higher test pattern count. But, it can effectively reduce the test power consumption in both the data path and clock tree. This is the preferred strategy for the capture power reduction.

Clock gating can be used to reduce the shift power with additional DFT hardware. In [49, 50], an independent clock control is associated with each scan chain that disables/enables the shift clock to drive the scan chains during scan shift. By shifting a portion of scan chains for each test pattern, the shift power consumed by both the clock tree and the combinational logic is reduced. The low shift power architecture shown in Figure 9.15 is extended to gate the shift clock in [51] such that a subset of the scan chains is enabled to be shifted in the test compression environment.

9.6.3 BLOCKING SCAN CELL OUTPUTS

Since the switching activity at the combinational logic in a design is caused by the state changes at the sequential state elements, the test power consumption in the data paths can be reduced by blocking the transitions that occur at the sequential state elements and propagate to the combinational logic during shift or capture. Two implementations of gating the scan cell output to be 0 and 1 are shown in Figure 9.18(a) and (b), where an AND gate and an OR gate are added as the blocking gates, respectively. By asserting the *Block_Enable* signal, the transition that occurs in the scan cells is blocked. During normal operation, this signal is deasserted.

To reduce the shift power, the method in [52] inserted the blocking gates at every scan cell output. In order to reduce the hardware overhead, a subset of scan cells, named power-sensitive scan cells, is identified by [53, 54] based on signal probability analysis, and the blocking gates are inserted only at these scan cells in order to achieve the trade-off between the shift power reduction and area overhead. Similar techniques based on the random simulation were proposed in [55] as well.

The method utilizing the blocking gates to reduce the capture power was pro reducing supply voltage droop to the scan chains in order to dynamically control the blocking gates during capture. The extension of applying this method in the test compression environment was also addressed in [56] by adding a capture power controller similar to the one shown in Figure 9.14.

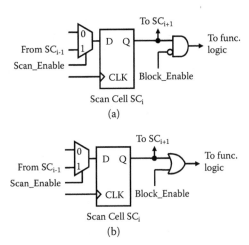

FIGURE 9.18 Gating the scan cell output: (a) with block value 0 and (b) with block value 1.

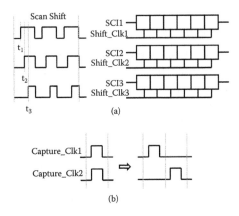

FIGURE 9.19 Clock application schemes: (a) shifting scan chains at different times and (b) applying clocks at different cycles.

9.6.4 CLOCK APPLICATION SCHEMES

Since the dynamic power consumption is triggered by clock pulse, the clocking schemes can be manipulated to reduce the test power with or without additional DFT hardware.

To reduce the shift power, the scan chains can be partitioned into multiple groups. Each group is either shifted at different times within a clock cycle [57] or shifted at the different clock cycles [58]. Figure 9.19(a) shows the clock scheme to shift three scan chains at different times, t_1, t_2, and t_3, in a shift cycle [57].

When doing test pattern generation, the test generator prefers to pulse as many clocks belonging to the same clock domain simultaneously as possible during capture in order to minimize test pattern count. In order to reduce the capture power, the ATPG tool can restrict the number of clocks pulsed in the same capture cycle.

To avoid the increase of test pattern count, those capture clocks belonging to the same clock domain are pulsed sequentially in different clock cycles, as shown in Figure 9.19(b). Since the test application time is dominated by the scan shift, adding extra capture cycles has negligible impact on overall test application time.

9.7 SOLUTIONS FOR REDUCING INDUCTIVE POWER SUPPLY NOISE

The low-power test solutions discussed in the previous section mainly address the resistant noise (IR-drop) issue by reducing test power consumption. Although the magnitude of the inductive noise, $L(di/dt)$, is decreased when the test power consumption becomes lower, those methods cannot eliminate the impact of the inductive noise. As described in Section 9.4.2, the inductive noise is triggered by abrupt change of circuit switching activity such that the power grid cannot deliver the power quickly enough to the cells with transition. In this section, we will describe several techniques to reduce the impact of the inductive noise on scan-based testing.

9.7.1 ON-DIE DROOP INDUCER

An on-die droop inducer is a large analog *field-effect transistor* (FET) that connects the power supply and the ground. Its gate input is controlled through scan configuration [30]. Since switching the FET on creates *di/dt* events that cause power supply droop, independently controlling the multiple on-die droop inducers spread across the design enables us to mimic the *di/dt* events with various magnitudes as well as make them happen at different die locations.

When applying LOC to do at-speed scan testing, there exists a time interval between the last shift cycle and the first capture cycle with very low or no switching activity. To avoid abruptly increasing the current demand when applying the first capture clock, the droop inducers are turned on during scan shift and turned off at the same time as the first capture clock becomes active, as shown in Figure 9.20 [30]. As a result, it avoids sudden change of the current by shifting the current from the droop inducers to the gates with switching activity. When turning on the droop inducers during shift, they should be switched on gradually to avoid breaking the shift operation due to the supply voltage droop.

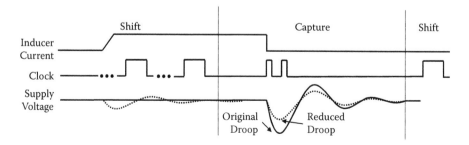

FIGURE 9.20   Reducing supply voltage droop by on-die droop inducers.

Besides reducing the supply voltage droop, the droop inducers can also be used to increase the droop when the test pattern has low switching activity that causes the voltage droop to be lower than the functional mode. On the other hand, if the power grid is fully charged right before applying the capture clock and the time interval for double pulsing the capture clock is much shorter than the resonant frequency of the first droop event shown in Figure 9.7, it is desirable to increase the voltage droop level to achieve better correlation between at-speed testing and functional operation. This can be realized by turning on the droop inducers simultaneously with the first capture clock pulse.

Although the droop inducers are very powerful to bridge the gap between at-speed testing and functional operation, it is worth considering the issues given below before deploying this technique:

- How many droop inducers should be inserted in the design and where should they be placed?
- The amount of reducing/increasing droop level is test pattern dependent. It has to be adjusted based on the switching activity of every test pattern.
- It is not a solution to deal with the voltage droop that fails the scan shift operation.

9.7.2 Scan Burst

To avoid sudden change of circuit state from a long period without activity to the capture mode, a warm-up clock sequence can be used to stabilize the power supply droop by increasing the circuit switching activity gradually during capture. A *burst-mode* clocking methodology was proposed in [31] to achieve this goal. In [31], a programmable burst clock controller is embedded in the design to control the shift clock frequency and the duration of the clock pulses. During capture, consecutive clock pulses are applied. Except the last two capture cycles, the clocks in some clock domains are slowed down by the burst clock controller, and they are pulsed every two, four, or eight cycles, such that the circuit is warmed up gradually. Figure 9.21 shows an example of burst-mode clock waveforms for an LOS test pattern.

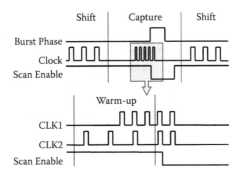

FIGURE 9.21 Example of clock burst waveforms. (From Nadeau-Dostie et al., Power-aware at-speed scan test methodology for circuit with synchronous clocks, In *Proceedings of International Test Conference*, 2008, Paper 9.3.)

In Figure 9.21, five consecutive clock pulses are applied in two clock domains, *CLK1* and *CLK2*, during capture. *CLK2* is run at half of its system speed in the first three cycles to warm up the circuit. To deassert the scan enable signal at-speed before the last clock pulse, a synchronous register driven by a slow-speed shift phase signal is used.

Since the clocks are operated at high frequency during warm-up, it may not be feasible to drive the scan chains from either automatic test equipment (ATE) or BIST controller directly. Therefore, the scan chains are arranged into circular segments based on clock domains, and each segment is required to be shifted by the same clock. During the regular shift cycle, the segments behave as standard scan chains to allow scan data to shift through. During capture, scan segments are operated in rotation mode except the last capture cycle, where the scan segment output feeds its input at very high speed. Since there is no data exchange between different clock domains, and between the clock domains and the ATE or BIST controller, the exact timing during the rotation phase is not critical. Only the timing for the last two capture cycles has to be set up accurately in order to detect delay faults at-speed.

Utilizing the scan burst helps to reduce the voltage droop on last capture. However, it requires pervasive changes of global design. For example, the scan paths have to be considered as at-speed paths, and the scan chains have to be configured as circular registers.

9.7.3   LATENCY BETWEEN SHIFT AND CAPTURE

Since the remnant current caused by the last shift clock pulse, I_{shift}, dies slowly and the instantaneous current, $I_{capture}$, is higher if the first capture clock pulse is applied after a long period without activity, the magnitude of $I_{capture}$ demanded by the first capture clock pulse can be reduced when the capture clock is applied before I_{shift} dies. Based on this assumption, the method proposed in [59] adjusts the latency between the last shift clock pulse and the first capture clock pulse in order to make the scan-based delay test frequency close to the functional frequency.

When testing a target function block, called tile, the latency between the last shift and the first capture is controlled by a programmable capture latency counter embedded in the tile. The counter is driven by the PLL clock. It releases the capture clock to the tile under test once it counts down to 0. Since the value of the capture latency counter must be no less than the routing delay to propagate the deasserted scan enable signal within the tile, physical optimization has to be done during the design phase in order to control the latency of the scan enable signal in every tile within an upper limit.

The experiment results given in [59] show that the method based on latency control is more effective at low supply voltage, but it needs to use different capture latencies for every tile in order to achieve the best improvement in capture frequency. Moreover, making neighborhood tiles quiet during capture helps to increase the capture frequency for the tile under test.

9.8 NEW TOPICS IN LOW-POWER TESTING

9.8.1 Toward Test Power Safety

Shrinking feature sizes and increasing clock frequencies have made timing-related defects a major cause for failing LSI circuits. This has made at-speed scan testing mandatory for achieving sufficient product quality. Compared with slow-speed scan testing, at-speed scan testing suffers from the risk of excessive test power [5, 6]. From the test safety point of view, excessive test power may cause two major problems: *heat-related test safety* (i.e., overheat may damage circuits) and *power-supply-noise-related test safety* (i.e., power supply noise may invalidate test responses).

Generally, heat-related test safety is determined by average shift power, which can be effectively and predictably reduced below a safety level by practical techniques, e.g., *scan chain segmentation* [60]. On the other hand, power-supply-noise-related test safety largely depends on the *launch switching activity* (LSA) caused by test stimulus launching at the beginning of the at-speed test cycle. As illustrated in Figure 9.22, based on the launch-off-capture (LOC) clocking scheme, the first capture, C_1, may cause excessive LSA, resulting in IR-drop and $L(di/dt)$ that reduce effective power supplies to cells, leading to increased path delay, and finally timing failures at the second capture, C_2. This section will focus on power-supply-noise-related test safety, referred to as *launch safety* hereafter, since launch switching activity (LSA) is its determining factor. Clearly, a launch-safe test vector is one that will not cause excessive LSA-induced timing failures.

Achieving launch safety requires sufficient reduction of excessive LSA. Recently, a scheme has been proposed for achieving guaranteed launch safety with minimal impact on test quality and test costs in power-aware test generation [61]. The basic idea consists of three integral parts:

1. **Risky path identification:** A *risky path P* of a test vector V is a path sensitized by V and has excessive launch switching activity (LSA) in its *impact area* (composed of the cells whose LSA impacts the delay of the path P). V is said to be *launch-risky* if it has at least one risky path.
2. **Risky path reduction:** Focused LSA reduction is conducted for the impact areas of risky paths in order to effectively reduce risky paths. This may turn a launch-risky test vector into a launch-safe one. Even if it cannot, it usually reduces the number of remaining risky paths.

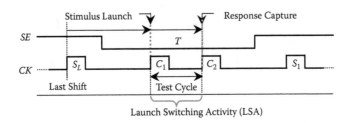

FIGURE 9.22 Launch safety in LOC-based at-speed scan testing.

3. **Risky path masking:** Since the value at the endpoint of any remaining risky path is uncertain due to excessive LSA in its impact area, it is excluded from use for fault detection. This is done by placing an X at the expected test-response-vector bit corresponding to that endpoint. As a result, no yield loss will occur. Note that this is data masking, without any performance penalty/additional circuit overhead.

Clearly, risky path masking is the core part to achieving guaranteed launch safety. Although being simple and straightforward, this part only becomes feasible under three critical conditions: (**CC1**) risky paths are identified, (**CC2**) the number of risky paths is small, and (**CC3**) an ATPG flow is devised to recover the lost fault detection capability due to masking. Only when **CC1** ~ **CC3** are all satisfied can risky path masking achieve guaranteed launch safety with minimal impact on test quality and test costs.

The scheme in [61] tries to satisfy the three critical conditions with a unique two-phase ATPG scheme. As illustrated in Figure 9.23, a test cube C_1 (with X-bits) is generated and then turned into a test vector V_1 (without X-bits) by detection-oriented X-filling (usually random-fill for high test quality and small vector count). Conventional ATPG ends here, but the new scheme continues with two more phases as follows:

Rescue: *LSP-based launch safety checking* (1) identifies all risky paths of V_1 by checking the LSA in the impact area of each *long sensitized path* (LSP) under V_1. Suppose that P_a and P_b are found to be risky paths. Then, *impact-X-bit restoring* (2) identifies those bits in V_1 that are originally X-bits in C_1 (before X-filling) and can reach the impact areas of P_a and P_b, and turns them back into X-bits (impact-X-bits) to create a new test cube C_2. After that, *focused low-LSA X-filling* (3) is conducted to turn C_2 into V_2 with reduced LSA in the impact areas of P_a and P_b.

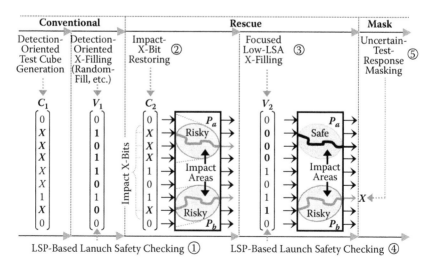

FIGURE 9.23 Basic idea for guaranteeing launch safety.

Mask: *LSP-based launch safety checking* (4) identifies that P_a is now safe but P_b is still risky under V_2. In this case, *uncertain-test-response masking* (5) is conducted to place an X at the endpoint (FF input) of P_b in the test response to V_2. This makes the uncertain value observed by the FF to be ignored in test response comparison, thus avoiding yield loss. Note that this masking needs no additional circuitry.

The advantages of the flow illustrated in Figure 9.23 are as follows:

Focused LSA reduction: LSA is reduced only for necessary vectors (launch-risky vectors) and only in necessary areas (impact areas). That is, there is no overreduction of LSA for launch-safe test vectors or in areas with low or timing-failure-non-causing LSA. This not only greatly improves the effectiveness of risky path reduction, but also avoids unnecessary test quality degradation.

Guaranteed launch safety: Masking any uncertain test response guarantees launch safety as the last resort. This is made possible by focused LSA reduction, which makes the number of remaining risky paths small, if any.

Minimal impact on test quality and test costs: Focused LSA reduction only uses necessary resources (i.e., impact-X-bits) but keeps original logic values at other bits already optimized by detection-oriented X-filling (e.g., random-fill). Furthermore, masking-induced loss in fault detection capability is mostly recovered by test vectors generated in subsequent ATPG runs. Therefore, the original test quality is preserved and severe test data inflation is avoided.

9.8.2 BASICS OF POWER-AWARE TESTING FOR 3D DEVICES

The three-dimensional (3D) technology offers a new solution to the increasing density of integrated circuits (ICs) [62, 75]. However, the new integration scheme also poses a set of challenges to design and test technologies. Meanwhile, heat dissipation in testing must be resolved before 3D ICs could become widely adopted. 3D ICs involve a complicated design and fabrication process. Accordingly, they are under a new set of constraints and challenges. Especially, testing and heat dissipation are of top priority. Without good solutions to these two problems, no 3D ICs would become feasible. The test problem for 3D ICs is unique because we have to test a partial system before the final layer of chip is assembled. On the other hand, the thermal issue is already a concern for 2D chips [77, 78], but it worsens in 3D ICs [63]. The major reason is that the upper chip layers, which are farther away from the substrate, have a less direct heat dissipation path. The excessive amount of heat would lead to poor performance and reliability. We consider placing a circuit to separate different layers like in [75], where each layer contains a subcircuit. Techniques in [65, 66, 72] cope with multicore/many-core SoC or NoC testing. Our method can establish a separate scan forest for each core.

A new thermal-driven scheme to reduce high-temperature hot spots incurred by the test process includes:

1. A scan tree architecture, which reduces the connection overhead and the number of TSVs, is established for postbond and prebond testing of 3D ICs. The architecture effectively reduces test application time and compresses test data.
2. A test ordering scheme is developed to avoid making the hot spots even hotter.
3. A new test application scheme is proposed to apply the ordered test set to the scan tree in the 3D IC under test so as to reduce the temperature produced by test application, which can still compress test stimulus data, compact test responses, and reduce test application cost. A lot of methods have been proposed to compress test data, compact test responses, and reduce test application time and test power by using new scan architectures [68–71, 73, 74, 79, 80]. A test compression scheme by an align-encode block was proposed to improve encodability of combinational test stimulus decompressors. Recently, a routing-driven scan tree synthesis method was proposed in [69] to compress test data, reduce test application time, and compact test responses. An interconnect-driven layout-aware multiple scan tree synthesis methodology for 3D ICs was proposed to reduce test data volume and test application cost in [8] most recently. New schemes to establish scan trees were proposed in [73] to compress test data, compact test responses, and decrease test application cost and test power. In a scan tree, each scan-in pin drives a couple of scan chains, where the scan flip-flops at the same level in the same tree are assigned the same value for all test vectors. Therefore, test data volume and test application time can be reduced significantly. A new scan architecture was proposed recently in [79] to reduce test response data by selectively collecting test responses in a low-power test compression environment. A fault is still detectable if only one of the scan flip-flops that receive the test responses of the fault is observable. It is found that test data can be well compressed. Test responses can be significantly compacted, and test power can be reduced greatly.

Initially, we have the netlist of the circuit. A circuit partitioning tool, *hMETIS* [67], is used to partition the circuit into multiple subcircuits, where it is better for any pair of subcircuits to have as few as possible interconnects. Each subcircuit is put to a layer of the 3D circuit. The minimum number of interconnects between any pair of subcircuits can introduce the minimum number of TSVs when establishing the scan trees for a 3D circuit. The new scan architecture is established to reduce the number of TSVs, like the technique in [72], where the good feature on test data compression and test cost reduction should still remain. The initial temperature information can be obtained by applying a small number of test vectors to the circuit after the scan tree has been established. The reason why we use scan tree for the thermal-driven scan testing is that we would like to use scan tree architecture to compress test stimulus data and reduce test application time for 3D circuits. Selecting a subset of test vectors is completed based on thermal information of the previous phase. Temperature of all nodes has been changed after each test vector. That is, hot spots may not be hot any more after some vectors have been applied, and some nodes that are not hot spots may become hot spots. The best way is to update thermal information after a vector has been applied; however, the CPU time for thermal analysis can be unacceptable. We must partition the test vector ordering process into multiple phases.

Figure 9.24 presents the test ordering scheme to reduce peak temperature in 3D ICs. The test ordering scheme is partitioned into multiple phases, where a subset of test vectors is selected in each phase. The selected test vectors produce the least power transition density on the hot spots, while it is not necessary to produce the least power transition density on all other nodes. Our method does not provide a globally

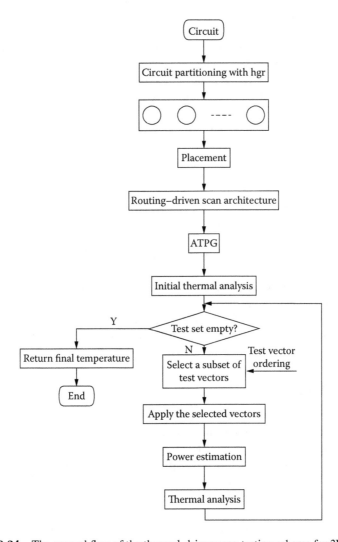

FIGURE 9.24 The general flow of the thermal-driven scan testing scheme for 3D ICs.

optimal solution for the problem, but a greedy procedure. Power consumption at the combinational part of the circuit is also included by using a cycle-accurate logic simulator. That is, each vector is selected to produce the least power transition density at the hot spots. The second vector is selected under the test responses of the first selected test vector. This process continues until the given number of test vectors has been selected. The thermal information of the circuit is updated after each subset of test vectors has been applied. It is not good to select too many vectors in each phase, which cannot provide accurate enough results for the thermal-driven test application scheme. The number of test vectors chosen in each phase also cannot be very small; otherwise, the CPU time to order tests and the CPU time to complete thermal analysis can be unacceptable. The process continues until all test vectors have been ordered. Actually, the ATE just has a single test phase with the ordered test vectors.

An existent thermal analysis tool, called ISAC2 [76], is used to estimate the temperature of ICs, which supports thermal analysis for both 2D and 3D ICs. A more popular analysis tool, HotSpot [64], can replace ISAC2. The 3D scan trees are established after the placement has been completed. The routing-driven scan tree architecture is established for test stimulus compression and test response compaction based on structural analysis, which minimizes the number of TSVs and the connection overhead. Our method gets the initial thermal analysis after the test vector set has been generated on the scan-tree-based 3D circuit and a few test vectors have been applied to the circuit. Our method selects a subset of test vectors based on the initial thermal analysis, which produces the lowest power transition density on the hot spots (gates or scan flip-flops with the highest temperature) during the period to apply the given number of test vectors. Thermal information of the circuit is updated after the selected test vectors have been applied by running the thermal analysis tool again, where power transition density, when applying the selected test vectors, is used for thermal analysis.

Another subset of test vectors is selected again based on the updated thermal information, which produces the least power consumption on the updated hot spots. The above process continues until all test vectors have been applied. The CPU time to run multiple rounds of the thermal analysis tool is shown to be much less than that of test generation and design for testability. The number of phases to select test vectors can be determined by a trade-off between accuracy of thermal analysis and the CPU time to complete test ordering, which is flexible enough.

9.9   SUMMARY

Low-power LSI circuits have become indispensable for almost all electronic applications, from battery-driven cell phones to energy-efficient microprocessors. However, it must be understood that achieving low functional power does not necessarily mean test power is also low. In fact, power dissipation can be much higher during LSI testing than during functional operations, resulting in severe test-induced yield loss due to heat damage or circuit malfunction. This has made low-power testing a must for low-power LSI circuits.

It is also important to understand that the new target of low test power must be added to the existing targets of LSI testing, i.e., high test quality and low test cost.

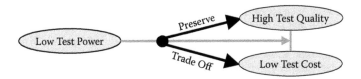

FIGURE 9.25 Holistic view of various targets in LSI testing.

As illustrated in Figure 9.25, test power should be handled properly in a holistic manner by carefully managing the impact of low-power testing on test quality and test costs. Generally, low-power testing must not degrade test quality but needs to trade off with test costs as optimally as possible.

The research and development for low-power testing is in a transitional phase now. Most current low-power testing solutions are coarse-grained, in that they can only achieve unfocused (global) test power reduction. Future low-power testing solutions may need to be fine-grained, in that they should achieve focused (pinpoint) test power reduction. The goals are avoiding overreduction, providing countermeasures in the case of underreduction, minimizing impact on test quality and test costs, and above all, guaranteeing test power safety instead of merely reducing test power.

In summary, efficient test power results in severe test-induced yield loss since it is much higher than functional power, especially for low-power circuits. For this reason, low-power testing must go hand in hand with low-power design. Only when this happens can a "cool" LSI product become a reality.

REFERENCES

1. International Technology Roadmap for Semiconductors, 2009, http://www.itrs.net/Links/2009ITRS/2009Chapters_2009Tables/2009_Design.pdf, 2009.
2. J. Rabaey, *Low Power Design Essentials (Integrated Circuits and Systems)*, Berlin: Springer, 2009.
3. Y. Zorian, A distributed BIST control scheme for complex VLSI devices, In *Proceedings of IEEE VLSI Test Symposium*, 1993, pp. 4–9.
4. S. Sde-Paz and E. Salomon, Frequency and power correlation between at-speed scan and functional tests, In *Proceedings of IEEE International Test Conference*, 2008, Paper 13.3.
5. P. Girard, Survey of low-power testing of VLSI circuits, *IEEE Design and Test of Computers*, 19(3): 82–92, 2002.
6. P. Girard, N. Nicolici, and X. Wen, eds., *Power-Aware Testing and Test Strategies for Low Power Devices*, New York: Springer, 2009.
7. M. Abramovici, M.A. Breuer, and A.D. Friedman, *Digital Systems Testing and Testable Design*, Piscataway, NJ: IEEE Press, 1994.
8. M.L. Bushnell and V.D. Agrawal, *Essentials of Electronic Testing for Digital, Memory and Mixed-Signal VLSI Circuits*, New York: Springer Science, 2000.
9. L.-T. Wang, C.-W. Wu, and X. Wen, eds., *VLSI Test Principles and Architectures: Design for Testability*, San Francisco: Morgan Kaufmann, 2006.
10. P. Girard, X. Wen, and N.A. Touba, Low-power testing, In *Advanced SOC Test Architectures—Towards Nanometer Designs*, L.-T. Wang, C. Stroud, and N.A. Touba (Eds.), San Francisco: Morgan Kaufmann, 2007, pp. 307–350.
11. C.P. Ravikumar, M. Hirech, and X. Wen, Test strategies for low-power devices, *Journal of Low Power Electronics*, 4(2): 127–138, 2008.

12. M. Tehranipoor and K.M. Butler, Power supply noise: A survey on effects and research, *IEEE Design and Test of Computers*, 27(2): 51–67, 2010.
13. B. Benware, C. Schuermyer, S. Ranganathan, R. Madge, P. Krishnamurthy, N. Tamarapalli, K.-H. Tsai, and J. Rajski, Impact of multiple-detect test patterns on product quality, In *Proceedings of IEEE International Test Conference*, 2003, pp. 1031–1040.
14. Y. Sato, S. Hamada, T. Maeda, A. Takatori, Y. Nozuyama, and S. Kajihara, Invisible delay quality—SDQM model lights up what could not be seen, In *Proceedings of IEEE International Test Conference*, 2005, Paper 47.1.
15. Z. Wang and K. Chakrabarty, Test-quality/cost optimization using output-deviation-based reordering of test patterns, *IEEE Transactions on TCAD*, 27(2): 352–365, 2008.
16. N.A. Touba, Survey of test vector compression techniques, *IEEE Design and Test Magazine*, 23(4): 294–303, 2006.
17. D.M. Walker and M.S. Hsiao, *Delay Testing, in System-on-Chip Test Architectures: Nanometer Design for Testability*, L.-T. Wang et al. (Eds.), San Francisco: Morgan Kaufmann, 2007.
18. X. Lin, et al., Timing-aware ATPG for high quality at-speed testing of small delay defects, In *Proceedings of ATS*, 2006, pp. 139–146.
19. S. Ravi, Power-aware test: Challenges and solutions, In *Proceedings of International Test Conference*, 2007, Lecture 2.2.
20. V. Chickermane, P. Gallagher, J. Sage, P. Yuan, and K. Chakravadhanula, A power-aware test methodology for multi-supply multi-voltage design, In *Proceedings of International Test Conference*, 2008, Paper 9.1.
21. C.P. Ravikumar, M. Hirech, and X. Wen, Test strategies for low power devices, In *Design, Automation and Test in Europe*, 2008, pp. 728–733.
22. K. Chakravadhanula, V. Chickermane, B. Keller, P. Gallagher, and S. Gregor, Test generation for state retention logic, In *Asian Test Symposium*, 2008, pp. 237–242.
23. K. Chakravadhanula, V. Chickermane, B. Keller, P. Gallagher, and A. Uzzaman, Why is conventional ATPG not sufficient for advanced low power designs? In *Asian Test Symposium*, 2009, pp. 295–300.
24. J. Rearick and R. Rodgers, Calibrating clock stretch during AC scan test, In *Proceedings of International Test Conference*, 2005, Paper 11.3.
25. J. Saxena, K.M. Butler, V.B. Jayaram, and S. Kundu, A case study of IR-drop in structured at-speed testing, In *Proceedings of International Test Conference*, 2003, pp. 1098–1104.
26. P. Pant and J. Zelman, Understanding power supply droop during at-speed scan testing, In *VLSI Test Symposium*, 2009, pp. 227–232.
27. A. Muhtaroglu, G. Taylor, and T. Rahal-Arabi, On-die droop detector for analog sensing of power supply noise, *IEEE Journal of Solid State Circuit*, 651–660, 2004.
28. H.Y. Loo, B.H. Oh, P.T. Oh, and E.K. Lee, CPU package design optimization for performance improvement and package cost reduction, In *International Conference on Electronic Materials and Packaging*, 2006.
29. R. Franch, P. Restle, N. James, W. Huott, J. Friedrich, R. Dixon, S. Weitzel, K. Van Goor, and G. Salem, On-chip timing uncertainty measurements on IBM microprocessors, In *Proceedings of International Test Conference*, 2007.
30. R. Petersen, P. Pant, P. Lopez, A. Barton, J. Ignowski, and D. Josephson, Voltage transient detection and induction for debug and test, In *Proceedings of International Test Conference*, 2009, Paper 8.2.
31. B. Nadeau-Dostie, K. Takeshita, and J.-F. Cote, Power-aware at-speed scan test methodology for circuit with synchronous clocks, In *Proceedings of International Test Conference*, 2008, Paper 9.3.
32. J. Wang, et al., Power supply noise in delay testing, In *Proceedings of International Test Conference*, 2006, Paper 17.3.

33. B. Kruseman, A. Majhi, and G. Gronthoud, On performance testing with path delay patterns, In *VLSI Test Symposium*, 2007, pp. 29–34.
34. P. Pant, J. Zelman, G. Colon-Bonet, J. Flint, and S. Yurash, Lessons from at-speed scan deployment on and Intel titanium microprocessor, In *Proceedings of International Test Conference*, 2010, Paper 18.1.
35. N. Ahmed, M. Tehranipoor, and V. Jayaram, Supply voltage noise aware ATPG for transition delay fault, In *VLSI Test Symposium*, 2007, pp. 179–186.
36. M.-F. Wu, H.-C. Pan, T.-H. Wang, J.-L. Huang, K.-H. Tsai, and W.-T. Cheng, Improved weight assignment for logic switching activity during at-speed test pattern generation, In *ASP-DAC*, 2010, pp. 493–498.
37. S.K. Goel, M. Meijer, and J. Pineda de Gyvez, Efficient testing and diagnosis of faulty power switches in SOCs, In *IET Computers and Digital Techniques*, 2007, pp. 230–236.
38. S. Khursheed, S. Yang, B.M. Al-Hashimi, and X. Huang, and D. Flynn, Improved DFT for testing power switches, In *European Test Symposium*, 2011, pp. 7–12.
39. H. Singh, K. Agrwal, D. Sylvester, and K.J. Nowka, Enhanced leakage reduction techniques using intermediate strength power gating, *IEEE Transactions on VLSI*, 15(11): 1212–1224, 2007.
40. Z. Zhang, X. Kavousianos, Y. Luo, Y. Tsiatouhas, and K. Chakrabarty, Signature analysis for testing, diagnosis, and repair of multi-mode power switches, In *European Test Symposium*, 2011, pp. 13–18.
41. K.M. Butler, J. Saxena, T. Fryars, G. Hetherington, A. Jain, and J. Lewis, Minimizing power consumption in scan testing: Pattern generation and DFT techniques, In *Proceedings of International Test Conference*, 2004, pp. 355–364.
42. D. Czysz, M. Kassab, X. Lin, G. Mrugalski, J. Rajski, and J. Tyszer, Low-power scan operation in test compression environment, *IEEE Transactions on CAD*, 28(11): 1742–1753, 2009.
43. D. Czysz, G. Mrugalski, J. Rajski, and J. Tyszer, New test data decompressor for low power applications, In *DAC*, 2007, pp. 539–544.
44. X. Wen, Y. Yamashitam, S. Kajihara, L.T. Wang, K.K. Saluja, and K. Kinoshita, On low-capture-power test generation for scan testing, In *Proceedings of VLSI Test Symposium*, 2005, pp. 265–270.
45. S. Remersaro, X. Lin, Z. Zhang, S.M. Reddy, I. Pomeranz, and J. Rajski, Preferred fill: A scalable method to reduce capture power for scan based designs, In *Proceedings of International Test Conference*, 2006, Paper 32.2.
46. X. Wen, K. Miyase, T. Suzuki, S. Kajihara, Y. Ohsumi, K.K. Saluja, Critical-path-aware X-filling for effective IR-drop reduction in at-speed scan testing, In *Proceedings of Design Automation Conference*, 2007, pp. 527–532.
47. W.-W. Hsieh, et al., A physical-location-aware X-filling method for IR-drop reduction in at-speed scan test, *IEEE Transactions on CAD*, 29(2): 289–298, 2010.
48. K. Chakravadhunula, V. Chickermane, B. Keller, P. Gallagher, and P. Narang, Capture power reduction using clock gating aware test generation, In *Proceedings of International Test Conference*, 2009, Paper 4.3.
49. R. Sankaralingam, M.A. Touba, and B. Pouya, Reducing power dissipation during test using scan chain disable, In *Proceedings of VLSI Test Symposium*, 2001, pp. 319–325.
50. C. Zoellin, H.-J. Wunderlich, N. Maeding, and J. Leenstra, BIST Power reduction using scan-chain disable in the cell processor, In *Proceedings of International Test Conference*, 2006, Paper 32.3.
51. J. Rajski, E.K. Moghaddam, and S.M. Reddy, Low power compression utilizing clock-gating, In *Proceedings of International Test Conference*, 2011, Paper 7.1.
52. S. Gerstendorfen and H.-J. Wunderlich, Minimized power consumption for scan-based BIST, In *Proceedings of International Test Conference*, 1999, pp. 77–84.

53. X. Lin and Y. Huang, Scan shift power reduction by freezing power sensitive scan cells, *Journal of Electronic Testing: Theory and Applications*, 24(4): 327–334, 2008.

54. X. Kavousianos, D. Bakalis, and D. Nikolos, Efficient partial scan cell gating for low-power scan-based testing, *ACM Transactions on Design Automation of Electronic Systems*, 14(2), 1–28, 2009.

55. M. Elshoukry and M. Tehranipoor, A critical-path-aware partial gating approach for test power reduction, *ACM Transactions on Design Automation of Electronic Systems*, 12(2): 1–22, 2007.

56. X. Lin and J. Rajski, Test power reduction by blocking scan cell outputs, In *Asian Test Symposium*, 2008, pp. 329–336.

57. T. Yoshida and M. Watati, A new approach for low-power scan testing, In *Proceedings of International Test Conference*, 2003, pp. 380–487.

58. S. Almukhaizim and O. Sinanoglu, Peak power reduction through dynamic partitioning of scan chains, In *Proceedings of International Test Conference*, 2008, Paper 9.2.

59. A. Majumdar, A. Sinha, N. Patel, R. Setty, Y. Dong, and S.-H. Chou, A novel mechanism for speed characterization during delay test, In *Proceedings of VLSI Test Symposium*, 2011, pp. 116–121.

60. L. Whetsel, Adapting scan architectures for low power operation, In *Proceedings of ITC*, 2000, pp. 863–872.

61. X. Wen, K. Enokimoto, K. Miyase, Y. Yamato, M. Kochte, S. Kajihara, P. Girard, and M. Tehranipoor, Power-aware test generation with guaranteed launch safety for at-speed scan testing, In *Proceedings of IEEE VLSI Test Symposium*, Dana Point, CA, May 2011, pp. 166–171.

62. P. Emma and E. Kursun, Opportunities and challenges for 3-D systems and their design, *IEEE Design and Test of Computers*, 26(5): 6–16, 2009.

63. Z. He, Z. Peng, P. Eles, P. Rosinger, and B.M. Al-Hashimi, Thermal-aware SoC test scheduling with test set partitioning and interleaving, *Journal of Electronic Testing: Theory and Application*, 24(1): 247–257, 2008.

64. W. Huang, M.R. Stan, K. Skadron, K. Sankaranayanan, and S. Ghosh, HotSpot: A compact thermal modeling method for CMOS VLSI systems, *IEEE Transactions on VLSI Systems*, 14(5): 501–513, 2006.

65. L. Jiang, L. Huang, and Q. Xu, Test architecture design and optimization for three-dimensional SoCs, In *Proceedings of Design Automation and Test in Europe*, 2009, pp. 220–225.

66. L. Jiang, Q. Xu, K. Chakrabarty, and T. Mak, Layout-driven test-architecture design and optimization for 3-D SoCs under pre-bond test-pin-count constraint, In *Proceedings of International Conference on Computer-Aided Design*, 2009, pp. 191–196.

67. G. Karyois, R. Aggarwal, V. Kumar, and S. Shekhar, Multilevel hypergraph partitioning: Applications in VLSI domain, *IEEE Transactions on VLSI Systems*, 7(1): 69–79, 1999.

68. H.H.S. Lee and K. Chakrabarty, Test challenges for 3-D integrated circuits, *IEEE Design and Test of Computers*, 26(5): 26–35, 2009.

69. K.S.M. Li, et al., Layout-aware multiple scan tree synthesis for 3-D SoCs, *IEEE Transactions on Computer-Aided Design*, 31(12): 1930–1934, 2012.

70. J. Li and D. Xiang, DFT optimization for pre-bond testing of 3D-SICs containing TSVs, In *Proceedings of International Conference on Computer Design*, 2010, pp. 474–479.

71. O. Sinanoglu, Scan architecture with align-encode, *IEEE Transactions on Computer-Aided Design* 27(12): 2303–2316, 2008.

72. X. Wu, P. Falkenstein, K. Chakrabarty, and Y. Xie, Scan-chain design and optimization for three-dimensional integrated circuits, *ACM Journal on Emerging Technologies in Computing Systems*, 5(2): 2009.

73. D. Xiang, K. Li, and H. Fujiwara, Reconfigured scan forest for test application cost, test data volume, test power reduction, *IEEE Transactions on Computers*, 56(4): 557–562, 2007.

74. D. Xiang and K.L. Shen, A thermal-driven test application scheme for pre-bond and post-bond scan testing of 3-dimensional ICs, *ACM Journal on Emerging Technologies of Computing Systems*, 10(2), article 18, February 2014.

75. Y. Xie, Y.J. Cong, and S. Sapatnekar, *Three-Dimensional Integrated Circuit Design: EDA, Design and Microarchitecture*, Berlin: Springer, 2009.

76. Y. Yang, Z. Gu, C. Zhu, R.P. Dick, and L. Shang, ISAC: Integrated space and time adaptive chip-package thermal analysis, *IEEE Transactions on Computer-Aided Design*, 30(1): 135–146, 2008.

77. Z. Chen and D. Xiang, A novel test application scheme for high transition fault coverage and low test cost, *IEEE Transactions on Computer-Aided Design*, 29(6): 966–976, 2010.

78. Z. Chen, K. Chakrabarty, and D. Xiang, MVP: Capture-power reduction with minimum violations partitioning for delay testing, In *Proceedings of International Conference on Computer-Aided Design*, 2010, pp. 149–154.

79. Z. Chen, D. Xiang, and B. Yin, The ATPG conflict-driven scheme for high transition fault coverage and low test cost," in *Proceedings of the VLSI Test Symposium*, pp. 146–151, 2009.

80. D. Xiang, D. Hu, Q. Xu, and A. Orailoglu, Low-power scan testing for test data compression using a routing-driven scan architecture, *IEEE Transactions on Computer-Aided Design*, 28: 1101–1105, 2009.

Index